T0074687

ENVIRONMENTAL DISASTER IN THE GULF SOUTH

The Natural World of the Gulf South

Craig E. Colten, Series Editor

.

ENVIRONMENTAL
DISASTER
IN THE
GULF SOUTH

TWO CENTURIES *of* CATASTROPHE,
RISK, and RESILIENCE

EDITED BY CINDY ERMUS
AFTERWORD *by* TED STEINBERG

Louisiana State University Press ▌▌▌ *Baton Rouge*

Published by Louisiana State University Press
Copyright © 2018 by Louisiana State University Press
All rights reserved
Manufactured in the United States of America
First printing

Designer: Michelle A. Neustrom
Typefaces: Sentinel, text; Brothers, display
Printer and binder: Sheridan Books, Inc.

"The Complete Story of the Galveston Horror: Trauma, History, and the Great Storm of 1900," by Andy Horowitz, was first published, in different form, in *Historical Reflections* 41, no. 3 (2015): 95–108, and is reprinted by permission of the editor.

Library of Congress Cataloging-in-Publication Data

Names: Ermus, Cindy, 1979– editor. | Steinberg, Theodore, 1961– writer of afterword.
Title: Environmental disaster in the Gulf South : two centuries of catastrophe, risk, and resilience / edited by Cindy Ermus ; afterword by Ted Steinberg.
Other titles: Natural world of the Gulf South.
Description: Baton Rouge : Louisiana State University Press, [2017] | Series: Natural world of the Gulf South | Includes index.
Identifiers: LCCN 2017005919| ISBN 978-0-8071-6710-6 (cloth : alk. paper) | ISBN 978-0-8071-6711-3 (pdf) | ISBN 978-0-8071-6712-0 (epub) | ISBN 978-0-8071-6713-7 (mobi)
Subjects: LCSH: Natural disasters—Gulf States—History. | Introduced organisms—Gulf States—History. | Emergency management—Gulf States.
Classification: LCC F296 .E58 2017 | DDC 363.340976—dc23
LC record available at https://lccn.loc.gov/2017005919

Para mi mamá y mi papá,
Elsa y Jorge Ermus

CONTENTS

1   Introduction
     CINDY ERMUS

14   Satire and Politics in the New Orleans
     Flood of 1849
     GREG O'BRIEN

37   Epidemics, Empire, and Eradication
     *Global Public Health and Yellow Fever Control
     in New Orleans*
     URMI ENGINEER WILLOUGHBY

62   The Complete Story of the Galveston Horror
     Trauma, History, and the Great Storm of 1900
     ANDY HOROWITZ

80   The 1928 Hurricane in Florida and the
     Wider Caribbean
     CHRISTOPHER M. CHURCH

103 Swamp Things
     *Invasive Species as Environmental Disaster
     in the Gulf South*
     ABRAHAM H. GIBSON *and* CINDY ERMUS

131 Political-Ecological Emergence of Space
     and Vulnerability in the Lower Ninth Ward,
     New Orleans
     ROBERTO E. BARRIOS

161   **Katrina Is Coming to Your City**
      Storm- and Flood-Defense Infrastructures
      in Risk Society
      KEVIN FOX GOTHAM

184   **Afterword**
      TED STEINBERG

195   **Contributors**

199   **Index**

ENVIRONMENTAL DISASTER IN THE GULF SOUTH

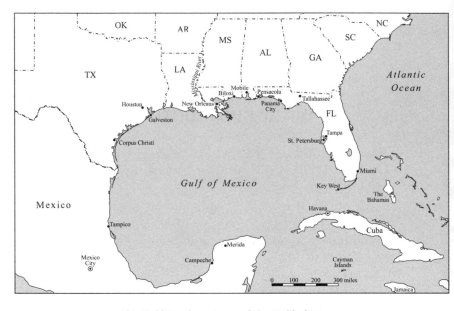

The Gulf South region and the Gulf of Mexico

Map by Mary Lee Eggart

INTRODUCTION

CINDY ERMUS

To understand the living present, and the promise of the future,
it is necessary to remember the past.
—Rachel Carson

Disasters represent an ever-present force in the human experience. They are instruments of history that have the power to shake the foundations upon which our societies and political systems are built, laying bare a region's social structures and vulnerabilities. As historians Elinor Accampo and Jeffrey H. Jackson have observed, "Disaster damage often highlights economic, racial/ethnic, and social power, an insight that comes as no surprise to American readers since Hurricane Katrina."[1] This has certainly been the case in the Gulf South region of the United States—a part of the world that has experienced more than its fair share of challenges brought on by an unforgiving natural environment on the one hand, and a history of political and commercial negligence on the other.

But can we say that this region—which encompasses Texas, Louisiana, Mississippi, Alabama, and Florida—is any more vulnerable in the face of natural hazards, climatic changes, and rampant mismanagement than some other parts of the world?[2] The Gulf South—especially the Gulf Coast—and the Greater Caribbean Basin to which it belongs is not unique in its propensity for disaster.[3] As Ted Steinberg has argued, Hurricane Sandy demonstrated that New York City is just as vulnerable, if not more so.[4] One might argue that the Pacific coast of the United States, susceptible not only to seismic activity and apocalyptic tales of "The Big One" but also to severe drought and forest fires, is the true home of disaster.[5] Perhaps it is Chile, which in 1960 experienced the world's most powerful earthquake (magnitude 9.5). Or is it the highly populated regions of Southeast Asia and the islands of the South Pacific, home to hundreds of active volcanoes and prone to frequent earthquakes? As Greg Bankoff has pointed out, "the Philippine archipelago . . . has the dubious distinction of rating the highest number of disasters of any during the twentieth century."[6] Japan, too, sits within a region of risk. The island country is vulnerable to earthquakes, tsunamis, and typhoons, and is home to 110 volcanoes, 47 of which are closely watched as a result of recent seismic activity.[7] Tellingly, Japan recently suffered the effects of what Sara B. Pritchard has called a "triple disaster," when a magnitude 9.0 earthquake struck the region on March 11, 2011, which resulted in a fourteen-meter tsunami, and the meltdown of half of the nuclear reactors at the Fukushima Daiichi nuclear power plant.[8] Wherever we look in the world, we are sure to find hotbeds of acute vulnerability with long histories of catastrophe. Today more than ever, as a result of our lightning-fast access to information from across the globe, we are confronted daily with news and images of earthquakes and tsunamis, epidemics, tornadoes, hurricanes, oil spills, droughts, fires, famines, and floods. And as the global population expands into more corners of the earth, increasingly building communities in areas that are vulnerable to the effects of climate change and natural hazards, our confrontations with these disasters and their threat will only increase.

For this reason, a closer look at the experience of the Gulf South with extreme events and risk is highly instructive. The vulnerabilities that have become characteristic of the region, especially since Hurricane Katrina, in many ways mirror the kind of challenges that pervade other parts of the world in times of crisis. The Gulf South may not be the only home of disaster, but few regions in North America are more often associated with catastrophe and risk. Over the last century alone, the Gulf South has suffered the effects of yellow fever outbreaks, major floods, catastrophic oil spills, significant droughts, the introduction of dangerous non-native species, as well as numerous devastating and record-breaking hurricanes. Meanwhile, it continues to struggle with the longer-term problems of rising seas, erosion, and disappearing wetlands, combined with the challenges posed by poverty, racial and ethnic tensions, and the vested interests of politicians, corporations, and private companies that operate in the background. As I write, in fact, parts of Louisiana are almost completely underwater, and a major rescue operation is underway in what Louisiana governor John Bel Edwards has called an "unprecedented and historic flooding event."[9] The correlation between events like this and climate change has not been lost on observers, from local witnesses, to academic researchers, to politicians, to Bill Nye the Science Guy, who question whether enough is being done to prevent such catastrophes.[10] In sum, as populations from all over the globe share in the experience of the not-so-slow disasters of climate change and political torpor, we can draw valuable lessons from the Gulf South's familiarity with natural hazards and social inequalities, which, in many ways, are applicable across time and space.

Yet the Gulf South, and the Gulf Coast in particular, is bound together by much more than geography or the shared experience of risk and vulnerability to wind, water, erosion, and biological exchanges. More fundamentally, the environment has helped define the region's identity and largely determined its history, its social fabric, and its economy.[11] Examples of environmental factors that characterize the Gulf South both in the popular imagination and in fact include its climate (heat and humidity), its water (lakes, rivers, bayous, swamps, and beaches), and its biota (mosquitoes, alligators, mangroves, live oak, Spanish moss, and a growing number of introduced non-native species that thrive in the accommodating environment). Indeed, writing about the Greater Louisiana region in particular, geographer Kent Mathewson has maintained, "the millennially framed workings and movements of streams, soils, seas, storms, and continental substrata may ultimately play a larger and more immediate role on this region than almost anywhere else on earth."[12] And never is the shared experience of the region as apparent as when a hurricane moves across South Florida to Texas or Louisiana, or when an oil spill devastates the waters, wildlife, and industry of the entire perimeter of the coast, as it did during the BP oil spill in 2010. Moreover, it is the environment of the Gulf South that has rendered it a tourist mecca, a retirement haven, and a home to the oil and seafood industries. People are drawn to the region's beautiful beaches and waterways, as well as to its economic and commercial activity, but as populations grow, so too do risk and vulnerability to extreme events.

Each essay in this volume examines a different moment or theme in the Gulf South's experience with disaster, risk, and vulnerability since the mid-nineteenth century. The pieces explore some of the social, cultural, political, economic, and ecological issues that surround the occurrence or threat of disaster from various disciplinary avenues, including history, sociology, and anthropology. These essays reveal not only how these cases differ but, in many respects, how they are the same. Most studies that have looked at vulnerability and disaster in the Gulf South have been published since 2005, and most have focused on the short- and/or long-term effects of Hurricane Katrina or the BP oil spill on the coasts of Louisiana and Mississippi.[13] Far fewer have been published on the broader Gulf South's experience with its unique and, at times, unforgiving environment prior to Hurricane Katrina, but this great storm does not represent the region's first encounter with the devastating effects of a natural hazard on an ill-prepared population.[14]

By looking at disaster and risk in the Gulf South across time and space, this volume provides a broad look at the variety of challenges that have plagued the region in times of crisis over the last century, as well as the ways in which these have been managed. Perhaps more importantly, in examining the underlying causes of our vulnerability to natural hazards, these studies suggest ways to increase resilience as we move forward, and underscore some of the lessons we can draw from past experiences—lessons that are applicable well beyond the Gulf South states, as societies around the world learn to cope with the effects of climate change and seek to build resilience and reduce vulnerability through enhanced awareness, adequate preparation, and efficient planning.[15] In other words, this volume not only looks at the problems, but also seeks to offer some solutions.

Problems and Contributions

Much as researchers from diverse disciplines are now acknowledging and/or debating the existence of the Anthropocene—that is to say, the current, but not yet formally accepted, geological epoch in which humans have become a dominant driving force in the environment—scholars have also increasingly documented the central role of human activity in creating and exacerbating the effects of disasters.[16] Especially since Hurricane Katrina, this idea has also gained more traction in the media and among the public. Numerous scholars, including Ted Steinberg, Craig Colten, Anthony Oliver-Smith, and contributors to this volume, have documented the role of politicians, vested corporate interests, deeply ingrained social inequality, and the neoliberal trend of deregulation and privatization that have become entrenched in our political systems over the past thirty-five years in creating the conditions that render communities especially vulnerable to environmental hazards.[17] Despite those like Pat Robertson who blamed the catastrophic 2010 earthquake in Haiti on a pact with the devil, and despite continued use of the phrase "Acts of God" in legal documents and insurance policies, generally speaking, more people than ever are aware of the human role in the occurrence of catastrophe—of the idea that, as historian Andy Horowitz put it, "the great storms may be inevitable, but the disasters that follow are not."[18]

This awareness, however, has not resulted in substantial change. Instead, we too easily forget the lessons learned from our past experiences with disaster. As historian Jean-Baptiste Fressoz has observed, it seems that, "The more disasters there are, the less we seem able to learn from them. Our faith in progress and our concern for economic efficiency make it clear that . . . we have not escaped from the illusions of modernity."[19] Historian Scott Knowles has referred to this as "national disaster amnesia," while geographers Craig Colten and Amy R. Sumpter refer to this as a loss of social memory.[20] Colten pointed out, "New generations of planners sometimes neglect the most obvious historical records and fail to adequately consider the complex human dimension when planning for hazards."[21] For example, twenty-three of the twenty-five most densely populated counties in the United States are on the coast.[22] Even though these coastal populations are especially vulnerable to rising sea levels and coastal erosion, development in these areas continues. Meanwhile, natural barriers along our coasts are destroyed in order to clear waterways or develop land. But not all hope is lost. As Knowles observed and as our volume will demonstrate, history has shown that this frustrating "disjunction between knowledge and actions" can be overcome.[23] Therefore, it is imperative that efforts to reveal the complex dynamics that surround the occurrence of, and vulnerability to, disasters are not slowed. These matters must continue to be given proper attention with ever more zeal in an effort to at least minimize our "loss of social memory," and in this way, to help reduce vulnerability and increase resilience in our communities. This interdisciplinary volume is an attempt to help fulfill this ongoing need. Its contributions proceed chronologically and represent various approaches to the study of disasters, risk, and vulnerability in the Gulf South.

In his study of the Mississippi River Flood of 1849 in New Orleans, historian Greg O'Brien reminds us that the challenges that confront the inhabitants of the Gulf South in times of crisis today are nothing new. Even in the mid-nineteenth century, the people of New Orleans rejected authorities' attempts to explain away the disastrous flood as an unpredictable act of God, and they did so publicly and with humor in local newspapers. Through biting sarcasm and witty anecdotes, citizens mocked local leaders, protesting their apparent lack of concern about the well-being and future of New Orleans, and their reluctance to approve necessary funding for flood prevention. In the end, this medium would represent a vehicle for change that led to significant reform and new projects in the city.

We find a different dynamic in Urmi Engineer Willoughby's essay on yellow fever eradication campaigns in New Orleans from the late nineteenth to the early twentieth century—a timely historical study as the Zika virus spreads through local transmission in South Florida. Situating the Gulf South within the larger scope of the Caribbean Basin, Willoughby describes early resistance to ecological understandings of the devastating yellow fever virus that explained its causes on the basis of urban growth and the presence of mosquitoes. Instead, for most of the nineteenth century, locals explained the illness as the inevitable consequence of filth, and therefore resisted efforts on the part of federal public-health officers to control its spread through the use of quarantines and other measures that locals feared would devastate commerce in the port city. By the turn of the century, however, cutting-edge research on the causes and transmission of yellow fever by prominent Cuban bacteriologists had finally reached American officials as a result of their increased presence in Cuba after the Spanish-American War. These new understandings of the virus allowed for the implementation of new measures that targeted mosquito breeding grounds, such as open containers of stagnant water, as the primary cause of the disease. After a final outbreak in New Orleans in 1905, the *vómito negro*, which had wreaked such havoc upon populations all over the globe in the eighteenth and nineteenth centuries, was finally eradicated in the United States.

Taken together, the essays in this volume encourage us to think of new ways of viewing and defining disaster. In his study of the Galveston storm of 1900, the deadliest and most disastrous hurricane in the history of the United States, historian Andy Horowitz, like O'Brien, reveals some of the ways in which societies and individuals try to make sense of the disorder that can follow a disastrous event. He reminds us that disasters do not merely represent sudden, ahistorical occurrences that arrive unexpectedly and end as soon as the storm clouds clear. Instead, the causes and effects of disasters transcend temporal boundaries, and thus represent ongoing experiences that warrant long-term observation. And the concept of trauma, as one of the longer-term manifestations of the effects of disaster, serves as a useful category of analysis for understanding disasters and their consequences. Horowitz applies this approach to the Galveston hurricane, exploring racism, violence, and inequality as inseparable from the disaster rather than as its consequences; the usefulness of his approach across time and place becomes clear as one considers the many parallels between the Galveston storm and subsequent disasters like Hurricane Katrina.

Historian Christopher M. Church also highlights disasters as long-term processes in his essay on the Okeechobee Hurricane of 1928—among the deadliest hurricanes in the history of the United States, second only to the Galveston storm. Like Willoughby, Church places the Gulf South within the Greater Caribbean Basin, tracing the storm's path of destruction as it hurtled toward Florida in an era when great agricultural and urban development had already begun to strip the Sunshine State of its natural barriers against the damaging forces of strong storms. As Roberto Barrios and Kevin Fox Gotham point out in their respective essays, the problem of disappearing coastal wetlands is ongoing in the Gulf South. And those who most suffer the burdens of vulnerability are society's disadvantaged—ethnic and racial minorities, women, and the poor. In the case of the 1928 hurricane, these were the laborers and migrant workers who had been transplanted to the region to work the vegetable and sugarcane fields and who lived in the least protected areas around Lake Okeechobee. Much as in 2005, the days following the storm would reveal the devastating results of the stark racial and social inequalities that remain woven into the social fabric of the Gulf South.

As the studies contained herein demonstrate, disasters and vulnerabilities can take many forms, and the deleterious effects of a hazard, much like their causes, can unfold over long periods of time. In their essay "Swamp Things," historians Abraham H. Gibson and Cindy Ermus explore the subject of invasive species, or more appropriately, introduced non-native species, within the context of disaster studies by looking at the long-term impact that pigs, nutria, fire ants, boll weevils, lionfish, and the Burmese python have had on the environments of the Gulf South. In doing so, Gibson and Ermus remind us that humans are complicit in the introduction of non-native species no less than they are in the exacerbation of a natural hazard's adverse effects on human populations. The effects of environmental crises like coastal erosion, rising sea levels, and the introduction of non-native species are not always immediately apparent, as they take place over longer periods of time than an earthquake or a hurricane, but this unawareness and inaction only add to their potential danger.

Meanwhile, no disaster in history was more televised and thus more visible to the public than Hurricane Katrina. And none has done more to garner the attention of scholars, journalists, and others who have felt compelled to help document the experiences of those whose voices had never previously been granted access to the historical record. The significance of this great storm cannot therefore be underestimated, for it has helped direct our attention to the countless ills that plague our most vulnerable societies both when threatened with disaster and in their aftermath. It has also forced us to examine the roles of our own social and political structures in exacerbating the effects of natural hazards. In their respective contributions to this volume, anthropologist Roberto E. Barrios and sociologist Kevin Fox Gotham each consider some of the lessons of Katrina, and demonstrate along the way the value in approaching disaster studies from various disciplinary avenues.

In his essay on space and vulnerability in New Orleans's Lower Ninth Ward after Katrina, Barrios reveals the kind of complex dynamics that exist on the ground in disaster-affected neighborhoods, and explores how these often-overlooked subtleties can affect disaster recovery and reconstruction efforts. Even well-intentioned attempts to help locals in times of crisis can be misguided and fall short of their promise if those who drive recovery efforts do not fully understand the concerns and inherent nuances that characterize local communities. Through his ethnographic research, Barrios was granted access into the experience of one of the most adversely affected neighborhoods in New Orleans during and after Hurricane Katrina. Through interactions with the inhabitants of the Lower Ninth Ward, he was able to identify local understandings of the environment, as well as divisions among residents that help drive local concerns. In the end, he reminds us that disaster recovery efforts are even more complex than they seem, and they should be approached as such.

In "Katrina Is Coming to Your City," Gotham examines urban infrastructures, including the failure of the New Orleans levee system during Hurricane Katrina, and demonstrates how decisions on the part of local and federal agencies, including decisions to underfund and impose budget cuts on infrastructure projects, can directly lead to disasters. In the case of the levee system in New Orleans, the U.S. Army Corps of Engineers neglected to incorporate acquired knowledge about storm surges, climate change, and coastal erosion into their project design in the years prior to Katrina, and the results were catastrophic. Gotham demonstrates that "risks do not happen by accident," nor do they exist as a result of natural or ecological processes. Instead, risks are directly linked to the decisions of individuals and/or organizations that fail to direct the resources necessary for increasing resilience and reducing vulnerability to potential hazards despite the financial risks posed to local and federal governments of such negligence or systematic failures. The title of his study is as fitting as it is instructive, for it reminds us that what happened in 2005 can happen again, and the Gulf South is by no means the only region at risk of such dangers.

The case studies in this volume offer new ways of understanding humanity's relationship with the environment of the Gulf South and beyond. Disasters must be remembered as ongoing crises that take many years to resolve, with causes that developed over long periods of time. The effects of Hurricane Katrina and the BP oil spill, for example, combined with the problems of erosion and the reshaping of southern Louisiana, have led to dramatically changing economic, geographic, and social landscapes that even now remain very much in flux, and there remains a great deal of uncertainty as to what the situation will look like in the years to come. Meanwhile, the ongoing recovery process in places like New Orleans has returned to being a local matter. The bright side is that scholars and members of the lay public are paying more attention than ever to the inexorable link between human decisions and the potential outcomes of a hazard, and this, we hope, is a precursor to positive change.

NOTES

1. Elinor Accampo and Jeffrey H. Jackson, "Introduction," *French Historical Studies* 36, no. 2 (Spring 2013): 165–66.

2. The Gulf South, in many ways, belongs to the much larger and interconnected world of the "Circum-Gulf and Caribbean realm," as Kent Mathewson has called it. By the same token, one could not overestimate the extent to which the Gulf South, most notably the Greater Louisiana region, interacted with and was shaped by the entire Mississippi River Basin that empties its waters into the Gulf of Mexico. "New Orleans and its environs were long the narrows that regulated passages cultural, commercial, demographic, and political between the Americas' Mediterranean and North America's heartlands" (Kent

Mathewson, "Greater Louisiana Connections and Conjunctures: Placing New Orleans in an Atlantic Time-Geographic Perspective," *Atlantic Studies* 5, no. 2 [2008]: 226).

3. The word "disaster" comes to us from the Greek and Latin terms for "bad star" or "star-crossed event." Today, the United Nations International Strategy for Disaster Reduction (UNISDR) defines the term as "a serious disruption of the functioning of a community or a society involving widespread human, material, economic or environmental losses and impacts, which exceeds the ability of the affected community or society to cope using its own resources." Meanwhile, a "hazard" is defined as "a dangerous phenomenon, substance, human activity or condition that may cause loss of life, injury or other health impacts, property damage, loss of livelihoods and services, social and economic disruption, or environmental damage." The distinction between these two terms is significant. As anthropologist Anthony Oliver-Smith and many since have pointed out, a disaster involves two elements, namely a natural or man-made phenomenon and a human population. A hurricane at sea, for example, only becomes a disaster when there are ships or human lives involved. See Anthony Oliver-Smith, "Introduction—Disaster Context and Causation: An Overview of Changing Perspectives in Disaster Research," *Natural Disasters and Cultural Responses,* ed. Anthony Oliver-Smith (Williamsburg: Department of Anthropology–College of William and Mary, 1986), 7.

4. Ted Steinberg, *Gotham Unbound: The Ecological History of Greater New York* (New York: Simon & Schuster, 2014).

5. For an example of this region's relationship with extreme events, in this case Los Angeles, see Mike Davis, *Ecology of Fear: Los Angeles and the Imagination of Disaster* (New York: Metropolitan Books, 1998).

6. Greg Bankoff, *Cultures of Disaster: Society and Natural Hazard in the Philippines* (London: RoutledgeCurzon, 2003), 4. See also Greg Bankoff, "Cultures of Disaster, Cultures of Coping: Hazard as a Frequent Life Experience in the Philippines," in *Natural Disasters, Cultural Responses: Case Studies Toward a Global Environmental History,* ed. Christof Mauch and Christian Pfister (Lanham, Md.: Lexington Books, 2009): 265–84.

7. David Cyranoski, "Why Japan Missed Volcano's Warning Signs," *Nature News,* September 29, 2014, www.nature.com/news/why-japan-missed-volcano-s-warning-signs 1.16000.

8. Sara B. Pritchard, "An Envirotechnical Disaster: Nature, Technology, and Politics at Fukushima," *Environmental History* 17 (April 2012): 220.

9. Debbie Elliott, "Unprecedented Flooding Batters Louisiana; Rescue Efforts Underway," National Public Radio, August 15, 2016, www.npr.org/2016/08/15/490037560 /unprecedented-flooding-batters-louisiana-rescue-efforts-underway (accessed August 15, 2016).

10. See, for instance, Chelsea Leu, "How Climate Change Will Redraw Louisiana's Flood Maps," *Wired,* August 17, 2016, www.wired.com/2016/08/climate-change-will-re draw-louisianas-flood-maps (accessed August 17, 2016); Chris Mooney, "What We Can Say about the Louisiana Floods and Climate Change," *Washington Post,* August 15, 2016,

www.washingtonpost.com/news/energy-environment/wp/2016/08/15/what-we-can
-say-about-the-louisiana-floods-and-climate-change/?utm_term=.d027dcfc141b (accessed August 17, 2016); Adam Epstein, "Bill Nye Explains That the Flooding in Louisiana Is the Result of Climate Change," *Quartz,* August 24, 2016, qz.com/765430/bill-nye
-louisiana-flooding-climate-change (accessed August 25, 2016).

   11. For more on the special character and history of what he calls the Creole Coast, see
Terry G. Jordan-Bychkov, "The Creole Coast: Homeland to Substrate," in *Homelands: A
Geography of Culture and Place across America,* ed. Richard L. Nostrand and Lawrence E.
Estaville (Baltimore: Johns Hopkins University Press, 2001): 73–82.

   12. Mathewson, "Greater Louisiana Connections and Conjunctures," 228.

   13. See, for instance, Michael Eric Dyson, *Come Hell or High Water: Hurricane Katrina and the Color of Disaster* (New York: Basic Civitas Books, 2005); Kent B. Germany,
"The Politics of Poverty and History: Racial Inequality and the Long Prelude to Katrina,"
Journal of American History 94, no. 3 (December 2007): 743–51; "Through the Eye of
Katrina: The Past as Prologue?" *Journal of American History,* special issue (December
2007): 743–51; Douglas Brinkley, *The Great Deluge: Hurricane Katrina, New Orleans, and
the Mississippi Gulf Coast* (New York: Harper Perennial, 2007); Amy Liu et al., eds., *Resil-
ience and Opportunity: Lessons from the U.S. Gulf Coast after Katrina and Rita* (Washing-
ton, D.C.: Brookings Institution Press, 2011); Vincanne Adams, *Markets of Sorrow, Labors
of Faith: New Orleans in the Wake of Katrina* (Durham, N.C.: Duke University Press, 2013);
Kevin Fox Gotham and Miriam Greenberg, *Crisis Cities: Disaster and Redevelopment in
New York and New Orleans* (New York: Oxford University Press, 2014); Craig E. Colten,
Southern Waters: The Limits to Abundance (Baton Rouge: Louisiana State University
Press, 2014).

   14. Some examples of studies that look at Gulf South disasters and/or vulnerabilities
prior to Katrina include: John M. Barry, *Rising Tide: The Great Mississippi Flood of 1927
and How it Changed America* (New York: Simon & Schuster, 1997); Erik Larson, *Isaac's
Storm: A Man, a Time, and the Deadliest Hurricane in History* (1999; New York: Vintage,
2000); Craig E. Colten, ed., *Transforming New Orleans and Its Environs: Centuries of
Change* (Pittsburgh: University of Pittsburgh Press, 2001); Craig E. Colten, *An Unnatural
Metropolis: Wresting New Orleans from Nature* (Baton Rouge: Louisiana State University
Press, 2005); Abby Sallenger, *Island in a Storm: A Rising Sea, a Vanishing Coast, and a
Nineteenth-Century Disaster that Warns of a Warmer World* (New York: Public Affairs,
2009); Cindy Ermus, "Reduced to Ashes: The Good Friday Fire of 1788 in Spanish Co-
lonial New Orleans," *Louisiana History* 54, no. 3 (Summer 2013): 292–331; Richard M.
Mizelle Jr., *Backwater Blues: The 1927 Mississippi River Flood in the African American
Imagination* (Minneapolis: University of Minnesota Press, 2014); Andy Horowitz, "Hur-
ricane Betsy and the Politics of Disaster in New Orleans' Lower Ninth Ward, 1965–1967,"
Journal of Southern History 80, no. 4 (November 2014): 893–934.

   15. The Community and Regional Resilience Initiative (CARRI), a federally funded
program that seeks to help communities put in place policies, practices, and processes
that will enable them to become more resilient in the face of hazard events, defines resil-

ience as "a community or region's capability to prepare for, respond to, and recover from significant multi-hazard threats with minimum damage to public safety and health, the economy, and national security." Meanwhile, vulnerability refers to "the potential for harm and social disruption from multihazard threats before hazard events occur." These concepts are very closely intertwined such that resilience can be increased by reducing vulnerability. See Craig E. Colten, Robert W. Kates, and Shirley B. Laska, "Three Years after Katrina: Lessons for Community Resilience," *Environment* 50, no. 5 (September–October 2008): 38–39.

16. More recent research has even observed the role of humans in creating the extreme event itself. See Adam H. Sobel et al., "Human Influence on Tropical Cyclone Intensity," *Science* 353 (July 2016): 242–46.

17. See, for example: Ted Steinberg, *Acts of God: The Unnatural History of Natural Disaster in America* (New York: Oxford University Press, 2000); Naomi Klein, *The Shock Doctrine: The Rise of Disaster Capitalism* (New York: Picador, 2007); Kevin Fox Gotham, "Disaster, Inc.: Privatization and Post-Katrina Rebuilding in New Orleans," *Perspectives on Politics* 10, no. 3 (Sept. 2012): 633–46; Anthony Oliver-Smith, "Conversations in Catastrophe: Neoliberalism and the Cultural Construction of Disaster Risk," *Cultures and Disasters: Understanding Cultural Framings in Disaster Risk Reduction*, ed. Fred Krüger, Greg Bankoff, Terry Cannon, Benedikt Orlowski, and E. Lisa F. Schipper (New York: Routledge, 2015), 37–52. See also the essays in Cedric Johnson, ed., *The Neoliberal Deluge: Hurricane Katrina, Late Capitalism, and the Remaking of New Orleans* (Minneapolis: University of Minnesota Press, 2011).

18. Horowitz, "Hurricane Betsy and the Politics of Disaster," 934.

19. Jean-Baptiste Fressoz, "The Lessons of Disasters: A Historical Critique of Postmodern Optimism," *Books and Ideas* (May 27, 2011), www.booksandideas.net/The-Lessons-of-Disasters.html (accessed August 1, 2016).

20. Craig Colten and Amy R. Sumpter, "Social Memory and Resilience in New Orleans," *Natural Hazards* (2009): 356; Scott Gabriel Knowles, *The Disaster Experts: Mastering Risk in Modern America* (Philadelphia: University of Pennsylvania Press, 2011), 20.

21. Colten and Sumpter, "Social Memory and Resilience in New Orleans," 362.

22. U.S. Climate Change Science Program and the Subcommittee on Global Change Research, *Coastal Sensitivity to Sea-Level Rise: A Focus on the Mid-Atlantic Region*, ed. James G. Titus et al. (Washington, D.C.: U.S. Environmental Protection Agency, 2009), 106.

23. Knowles, *The Disaster Experts*, 20.

SATIRE AND POLITICS IN THE NEW ORLEANS FLOOD OF 1849

GREG O'BRIEN

Respectfully Dedicated to the "Authorities of New Orleans"

Oh! Mr. Mayor, how can you sleep
When the water in town is nearly knee deep,
And tadpoles and frogs are playing about,
As if their dear mothers "didn't know they were out."

Why did you let this big crevasse come down,
To pay a short visit to its friends here in town,
With its "cortege" of polliwogs, fishes, and snakes;
Its sister "the fever"—its brother "the shakes."

Our Aldermen, too—the most learned group
That ever did justice to mock-turtle soup,
Are now of the opinion, that till something is done,
The flood from the river will continue to run.
—"City Lyrios," *Daily Crescent* (New Orleans), May 28, 1849

The Setting

New Orleans occupies a dominant space in contemporary American consciousness when it comes to flooding and the role of government, engineering, and culture in responding to natural disasters. That was no less true in 1849 when the entire nation's newspaper media and even the foreign press carried daily updates in May and June about the latest flood to strike the Crescent City. On May 3, 1849, the Mississippi River broke through the levee at Providence plantation, belonging to Pierre Sauvé, seventeen miles upriver from the center of New Orleans. Since the city's founding in the early eighteenth century, man-made levees along the river had gradually increased the height of the natural levees as the only protection against river floods, but levee design and construction fell to the private landowners outside city limits, and quality of construction remained uneven. By May 8, waters had begun reaching the back parts of the city's Second Municipality. Within days, the river flooded about 220 occupied city squares, surrounded two thousand tenements with water, forced nearly twelve thousand persons to flee their homes, and covered some cemeteries with waters as deep as six feet. The human and economic toll was staggering. Some residents died in or near their homes, tens of thousands of the city's poorest residents were displaced, and inhabitants threatened one another with violence over which parts of the city should be protected and which parts sacrificed. Food shortages and lack of clean drinking water exacerbated an already miserable situation. Snakes and alligators patrolled the swamped streets and buildings, or at least were rumored to have done so.[1]

The city's response to the flood quickly became a political scandal as elected officials dithered over appropriate responses to the many crises that erupted as floodwaters poured into the city. Overwhelmed by the crisis, city government and engineers could not agree on the best strategy to protect citizens or fix the levees. Engineers blamed politicians for ignoring sound scientific advice and refusing to fund the necessary protection measures, while politicians blamed engineers and other politicians for the muddled response. Some of the city's residents used the social media of their day—the newspapers—to anonymously berate their political and socioeconomic leaders for their insufficient response to the levee breach and ensuing flood. Invoking humor and satire in original poems, fiction, and limericks, New Orleanians demanded more from their elected officials in the face of this unprecedented catastrophe, and this nineteenth-century rhetorical device reveals much about the political response to this flood, as well as citizen demands for better handling of natural disasters.[2]

As historians, geographers, and environmental scientists have been arguing for some years now, there is no such thing as a "natural disaster." An environmental event such as a flood, tornado, hurricane, or earthquake only becomes disastrous when it has a negative impact on human life and property; otherwise, it is merely the proverbial tree in the forest that is deemed to have insignificant impact because no humans are affected. Environmental historian Ted Steinberg demonstrated, for example, that business and political interests began trying to normalize natural disasters in the late nineteenth- and early twentieth-century United States in order to quickly restore order and economic growth in areas hit by natural hazards. Natural disasters began to be seen less as the work of a vengeful God and more as unpredictable and unpreventable acts of nature. Thus, residents of areas prone to environmental problems shouldered the bulk of the responsibility for their own recovery after a natural disaster. By turning the focus away from the human role in creating natural disasters, Steinberg maintains, these economic and political leaders removed their responsibility for the impacts of natural hazards.[3]

The recent scholarly attention to natural disasters has highlighted the fundamental paradox within the United States of an economy based upon constant development that encourages human settlement of areas prone to environmental problems. Today it is up to engineers to make such lifestyles possible and safe, just as in 1849 it was up to city and state engineers, as well as plantation owners, to maintain and strengthen the flood-protection system along the lower Mississippi River. No matter how good the engineering science, though, significant sums of money had to be spent in order to actualize engineering plans, which means that politics played as much a role in municipal civil engineering as science, as was the case in New Orleans in 1849. The history of the New Orleans Flood of 1849 must therefore explain the politics as much as the engineering and the science and the weather in order to illuminate all of the factors at work in creating this "natural" disaster. The newspapers of the day make clear that citizens in antebellum New Orleans realized the intimate connection between politics and engineering and that they thought human actions could have prevented the crevasse and lessened the impact of the flood once it occurred.

No American city has had to concern itself with water the way that New Orleans has. The city has always been prone to flooding from the Mississippi River due to high water, from Lake Pontchartrain due to storm surge from hurricanes or high winds pushing the lake water onshore, and from rain that collects in the low-lying areas of the city faster than gravity or pumps can get it out. New Orleans's unique geography means that the highest elevations in the city are on the natural and man-made levees along the Mississippi River, with a natural bowl lying below sea level between the river levees and Lake Pontchartrain. In the mid-nineteenth century, this low-lying bowl was hemmed in by the river to the south and the Metairie Ridge to the north—the natural levee remains of a former path of the Mississippi River. The floodwaters rushing through Pierre Sauvé's broken levee in 1849 flowed east between the river levee and the Metairie Ridge, directly into the upriver and back parts of the Crescent City.

Somewhat surprisingly, the attempt of some New Orleans officials to portray the 1849 disaster as unforeseen and due to the folly of nature fell flat among the city's media and citizenry. Even though the roots of the booster mentality can be found in the official response to the flood, whereby residents were told to buck up and deal with the circumstances as best they could, a louder reaction by engineers, newspaper editors, and citizens insisted that the government should do everything in its power to solve the crisis rather than sit by idly while nature ran its course. City newspapers asked, "Who Is Responsible?" and described the city's response to the flood as "Gross Imbecility."[4] Several newspapers suggested that corruption or at least uncaring mismanagement hampered efforts to ameliorate the suffering of thousands of the city's residents. While public officials argued with each other and engineers scrambled to find solutions, citizens used satire to insist that action be taken. Their mockery of city officials forced city leaders to respond, even if only to defend their inaction.

The Politics of the Flood

By the 1840s the city and its three municipalities maintained the levees within city limits, but the towns of Lafayette and Carrollton (now within New Orleans) and private landowners, especially plantation owners, maintained the levees upriver from New Orleans. Those levees outside city limits were subject to inspection, however, on orders from the New Orleans mayor, A. D. Crossman. After an unusually warm and wet winter in 1848–49, the Mississippi River rose at least fifteen feet above normal in the early spring to heights that brought the river to the tops of the levees and that even longtime residents could not remember seeing before. In late February 1849, Mayor Crossman ordered engineers to inspect the levees upriver from New Orleans to assess the danger to the city from the high river. Into this urgent situation waded George Towers Dunbar Jr. A Maryland native, Dunbar had worked as a civil engineer and surveyor for several railroad companies, eventually moving to New Orleans in 1835 while working for the New Orleans and Nashville Railroad, whose construction above New Orleans ended with the nationwide financial panic of 1837. From 1838 to 1843, he served as engineer of the Board of Public Works of the State of Louisiana and worked on several projects, including the sewage-disposal problem in New Orleans and the removal of obstacles at the mouth of the Red River. Dunbar was removed from the state-engineer position in 1843 when the new Democratic governor, Alexandre Mouton, exercised political patronage and removed Whigs, such as Dunbar, from state office. Later that year, Dunbar was appointed assistant surveyor of the Second Municipality of New Orleans and then selected as the primary surveyor in 1844. In 1836, the Louisiana legislature had divided New Orleans into three municipalities, with the Second Municipality comprised of areas upriver from Canal Street, where the newer American and English-speaking population had congregated (approximately today's Central Business District). The surveyor's job included surveying property boundaries, keeping streets clean, repairing damaged streets, contracting for street and municipal building construction, designing drainage systems, and maintaining the levees in the municipality.[5]

Dunbar issued an alarming report. The height of the river was sixteen inches above the high-water mark of the 1816 flood, the previous worst flood in New Orleans's history. He found the levee just upriver from New Orleans near Carrollton in a state of extreme danger as the Mississippi River had eaten away the base of the levee while residents in the area foolishly transferred dirt from the land side of the levee to the top to raise the height, thus weakening the levee's stability still further. Dunbar then took the opportunity to chastise the city's political leaders for their lack of planning in protecting the city from floods:

Annually, when ruin has stared us in the face, the Mayor has called upon the Surveyors to report upon the Carrollton levee, and they have invariably reported it dangerous in the extreme; but still no crevasse has taken place, and, in consequence, nothing has been done by the authorities to avert so awful an evil. And so it will be. The Councils will go on dreaming of their security, until they find one half of the city and suburbs engulphed, and millions destroyed. They will then regret that ample means were not taken for our security. I feel it my duty, sir, to plainly state that the authorities of this city, Lafayette and other suburbs are trifling with the property and lives of our citizens; and, sir, although I wish to create no panic, or cause unnecessary alarm, I say, the time is not far distant when they will regret it. It may be years before a crevasse will take place, and it may occur in a few days—the latter the more probable.[6]

Dunbar proposed both short-term and long-term solutions to the threat of Mississippi River floods. He called for the immediate construction of new levees between the river and the Metairie Ridge upriver from the inhabited areas of the city to direct any floodwaters toward the swamps on the lake side of the Metairie Ridge. For more permanent protection from the river, he insisted that a mile and a half of Mississippi River frontage be purchased many miles upriver from New Orleans and a large drainage canal be constructed from the river to Lake Pontchartrain to divert excess river water away from the city at times of high water. Exasperated, he asked the Second Municipality Council, "What is the cost compared with the safety of our lives and property[?]" "Put it to a vote," he added, "and this city will bear cheerfully the expense."[7] Unfortunately, the issue was never put to a vote.

In fact, the chairman of the Second Municipality Council, the Whig banker and real estate developer Samuel Peters, reacted with repugnance to Dunbar's proposals and tone. He admonished Dunbar for presuming to know the political temperament of the people of New Orleans and stated flatly that "it was preposterous" to suppose that they would agree to the expense of a new canal connecting the Mississippi River with Lake Pontchartrain. Peters seems to have been more upset by Dunbar's lack of deference, however, than by his specific proposals. Dunbar's "communication was disrespectful," warned Peters, and he "should have confined himself to his duties ... stated facts, and given such professional advice as the occasion called for, and not take upon himself to censure the authorities."[8] The only resolution passed in early March by the Second Municipality Council in response to Dunbar's warning was that the surveyor should be supplied with laborers and carts "to meet any emergency that may arise."[9] On March 3, however, the mayor's office took Dunbar's report more seriously and convened a conference committee made up of representatives of the three municipal councils, the cities of Lafayette and Carrollton, and Jefferson Parish that adopted a short-term solution to build a temporary inner levee behind the weakened one at Carrollton. Their action did not take into account that the levee might break somewhere other than Carrollton, which it did on May 3.[10]

Satire and Criticism

Less than two weeks after the levee failed on Sauvé's plantation, the *Daily Delta* newspaper savagely attacked the legislative inaction of Alderman Peters and reprinted the dispute between Dunbar and Peters over how to prepare for the flood. Published opinion in the city mostly sided with Dunbar, accusing the Second Municipality Council of negligence and Alderman Peters in particular of having more concern for paying interest on municipal bonds than on spending money to protect citizens and their property.[11] A large-scale public relations battle ensued with papers such as the *Commercial Bulletin* defending Peters while the *Daily Delta,* the *Weekly Delta,* the *Daily Crescent,* and the *Daily Picayune* sided with Dunbar. Perhaps fearing the political fallout as the public began to take a hard look at his lack of action, Peters attempted to change the public's perception of Dunbar. On May 13, Peters sent an out-of-town engineer named Keating to visit the crevasse and publish a letter in the *Daily Crescent* deeply critical of Dunbar's efforts to direct the closing of the crevasse.[12] Thus, on May 17 an exhausted Dunbar journeyed back from Sauvé's plantation to New Orleans to hold hearings with the mayor and the newly formed Joint Committee on the Crevasse, composed of officials from the three municipalities, including Peters. The crevasse was not yet fixed, and the Crevasse Committee, at Peters's insistence, called in a Captain Grant of Mobile, who had apparent "practical" experience in closing crevasses, to take command of repair efforts.[13]

Within days of Peters's exercise of political power, several New Orleans newspapers began publishing humorous yet biting criticism of the city's response to the flood crisis. As it became obvious that the crevasse would not soon be closed, engineers and people claiming to have engineering expertise put forward various ideas. Less than two weeks after floodwaters began reaching the city, one satirist writing for the *Weekly Delta* hit upon a simple colloquially worded solution amid all of the engineering chatter:

I shuddent be stonished et the yung fellers bout toun shood rase a fuss with the Mayer and Aldermen of the citty fur not stopin the crevass, cause its knok'd thare Sundy drivin on the Shell rode into next summer sum time; and ef thay shood take a noshun thay'll vote all them fellers out of offis. Its reely laffable to here and rede of the difrent ways evryboddy has got to stop it. Sum or yure frends rites won way, an sum of the Picayune's fends anuther, an yesterday morning the Crescent wanted'em to send up the Stait Ingeneer, and the Bulertin ses in the number of councilors thare is wisdom, an is goin to giv us a raft for diffrent ways to stop it. Thay puts me in mind of a feller as is sik, an insted of hain a fust rate doctor rite off, he takes fust won thing an then anuther, an keeps changin so oftin, that he gets wuss an wusser, an bimeby dies. My privit pinion bout it is, that the best way to stop it is to put sumthin in to kepe the water frum runnin throo, an I dont believe thay'll git it stopt till thay do that.

Yure frend,   Cap'n Bender.[14]

"Cap'n Bender's" idea to put *something* in the crevasse to keep the water from running through expressed the frustration of many residents that city officials should stop talking about what to do and instead take action. Another writer, "The Man on the Bridge," was so taken by "Cap'n Bender's" suggestion that he repeated it in the New York weekly *Spirit of the Times,* adding that the "principal 'Taing about Town' now, is water, which covers about one-fourth of our city."[15] At this time, the *Daily Delta* also poked fun at the circumstances that people in the flooded areas had to endure by publishing a supposed "Dialogue Between Two Residents of the Submerged District": "'Well, Dave, how do you get along in the Second [district]?' 'Oh, *swimmingly.* But tell me, Bill, how are you off in the old Seventh?' *Bill—*'Barely able, my dear fellow, to keep our heads above water.'"[16]

New Orleans has long been justifiably recognized for its contributions to nineteenth-century American humor. By the 1840s, the city's newspapers drove the publication of tall tales, embellished scenes of city life, and other humorous asides. Newspapers throughout the country published similar takes on local life, but New Orleans was a rapidly growing and uniquely multicultural city in antebellum America. Sitting near the mouth of the massive Mississippi River system, the city seemed destined for greatness during the market revolution sweeping the country in the antebellum era. As Americans and foreigners moved there by the tens of thousands in the 1830s and 1840s, they were in turns fascinated and repelled by the customs, languages, and lifestyles of New Orleans. The city's better-established French, Spanish, free black, and slave populations were joined by Americans and slaves from other regions and large numbers of German-speaking and Irish peoples in the 1840s. Published humor became a way to criticize or celebrate local cultural quirks, and to emphasize that "life is different here." Besides the several city newspapers, the popular New York–based weekly *Spirit of the Times* published New Orleans writers. William T. Porter, editor of the New York paper, was a close friend of George Wilkins Kendall, editor of the *New Orleans Picayune,* and two of Porter's brothers worked for the *Picayune* for a stretch. In 1846, a collection of humorous character sketches that had originally appeared in the *Picayune* was published by a Philadelphia printer, bringing this unique brand of humor to a national audience and forever connecting the city with ribald tales of absurdity and comedy.[17]

Even before 1849, the environmental vulnerability of New Orleans provided material for humorists. As a city built on the sinking soils of the Mississippi River delta and below sea level, New Orleans remained particularly prone to flooding and preserved relative safety only with engineering intervention. But New Orleans was also a gateway to the Mississippi, Arkansas, Missouri, and Ohio river valleys, among others, and thousands of people stopped in New Orleans in the antebellum era just briefly before "floating" upriver via steamboat. In "Notes for a Tourist" in 1841, an anonymous contributor to the *Picayune* played upon the perception that New Orleans was under constant threat of flooding: "You will hear, Madam, a great deal about the '*floating* population of New Orleans,'—a phrase which you will understand when I tell you that the town is half the year under inundation from the Mississippi."[18]

Making light of the environmental vulnerability of New Orleans may not have been new in 1849, but the use of satirical humor to directly criticize the actions of city government and to call for more attentive behavior from elected officials repurposed humor towards political ends. Satire provided an effective means to critique public officials by suggesting that their actions or ideas should be viewed with disdain and that alternative policies should be pursued. Political satire often employed irony to criticize official behavior as ineffective if not damaging to public well-being. Distinct from mere humor, powerful political satire leaves the reader viewing public officials or their policies as absurd, and demanding that more effective persons, ideas, and actions be adopted. In nineteenth-century and earlier European and American history, satire became particularly prominent at moments of societal crisis, such as war, economic depression, political oppression, or, as argued here, natural disasters. New Orleanians in 1849 could draw upon a rich tradition of well-known English-language satirists such as Washington Irving, Lord Byron, Jonathan Swift, and Charles Dickens. The rich newspaper and publishing culture of mid-nineteenth-century New Orleans provided the forum for a significant literate and well-read city population to vent satirical displeasure during their city's time of inundation.[19]

Anonymous writers "flooded" newspaper editorial offices "with poetical effusions," leading the editor of the *Daily Delta* to select from "these postdiluvian ditties" a poem entitled "The Crevasse, or No Crevasse!" playing on the Shakespearean "To be, or not to be," soliloquy, and calling on the city to relieve the sufferings of the Second Municipality's wards by cutting through the Metairie and other ridges to funnel the floodwaters towards Lake Pontchartrain. Published on May 26, the poem read in part: "The Crevasse, or No Crevasse! that is the question. / Whether 'tis nobler in the Crescent City to suffer / The flow of Father Mississippi's mighty waters / Through that small crevasse of three hundred yards, / Or, by opposing, stop it?"[20]

While the flood deepened and spread throughout the city in late May, the issue of who was to blame never left the minds of city residents. Writing from New Orleans on May 26, the editor of the Natchez-based *Mississippi Free Trader and Natchez Gazette* stated flatly that the issue of blame consumed the general topic of conversation in the city. "Some aver that Mr. Peters is the gentleman [to blame] because he opposed the valuable suggestion of the engineer, Mr. Dunbar," he wrote, "while others affirm that the Mayor, A. D. Crossman, is the only man to blame, as he is the chief officer here and is liberally compensated for attending to, and to give timely notice of any damage which might threaten, or endanger the welfare of the city."[21] In the same issue, under a column titled "Who is Responsible?" the *Daily Delta* editors condemned certain city officials, such as Peters, for favoring so-called "practical" skills in fixing levees over scientific and professional expertise and for blaming "scientific men," such as Dunbar, for failing to solve the crisis. Such ignorance of science, they wrote, "as if true science was anything more than the perfection of practice—the results of experience—the rules collected from a close observation of the workings of nature," provided further fodder for questioning the authority of city political leaders.[22]

Tone deafness to the plight of the city's residents and to the future well-being of New Orleans on the part of the city's elected officials and economic leaders provided the primary point of criticism as the flood persisted into late May and early June. Few critiques of city governance reached the nearly revolutionary warning published in the *Daily Delta:* "If the people do not take matters in hand, the interests of our city will be seriously compromised. If our timid, unenergetic, indifferent property-holders, our useless rich men and capitalists, do not arouse from their shameful inactivity, and take some part and responsibility in the management of our corporation affairs, the vast resources and blessings which a kind Providence has bestowed upon us, will be insufficient to prevent the decline of our city's wealth and prosperity."[23] Such class-based criticism was rarely so bluntly published in the city's papers, but exasperation at the city's tepid response to the flood could be felt everywhere.

Aldermen of the Second Municipality Council had taken only one major action by late May to aid those affected by the flood: they provided skiffs and other boats to citizens in the flooded areas, though only at some cost to those who wanted to use the skiffs' services. The editors of the *Daily Crescent* thought, however, that "something more should have been given than mere transportation." The boats were a convenience, but "many poor people, who are driven from their homes by the flood, know not where to go or what to do, when they get to dry land."[24] When some members of the Second Municipality Council proposed cash payments to the poor in the municipality to help them acquire basic supplies and housing, Alderman Peters struck down the idea as "charity" and insisted that private sources would pick up the slack.[25]

Many New Orleanians did not care for the attitude of Peters and other like-minded officials toward their misfortune. Grocer E. J. White delivered groceries by boat to his customers' homes during the flood without charging extra for the service. He refused to extend the same courtesy to the aldermen of the Second Municipality, however. "Mr. White, like a thousand others of the *floating* population, would no doubt cheerfully take all the city authorities in his boat for one trip at least," the *Daily Delta* reported, "were it only for the satisfaction of dumping them out of soundings, in the deepest water of the inundated district."[26] The refusal of Second Municipality officials to provide public relief for victims of the flood inspired one anonymous author to pen new lyrics to Charles Dickens's satirical song "Fine Old English Gentleman," originally written in 1841 to criticize the new conservative Tory government in Britain and its blinders to the suffering of average people. This ditty's lyricist insisted that "the majority of the property holders and residents of the Seventh Ward [in the Second Municipality] will agree with me as to the truth of the complaints contained in [this] rigmarole":

I.

There was a jolly Alderman down in the Seventh Ward,
Who, when his *dry* constituents were from dry land debarred,
Got up among the council-men—strong proof of his regard,
And damned the skiffs, and said the sound upon his ear fell hard.
Oh, it was so unmusical! It would not do at all.

II.

But strange to say from that night out, no change at all, was seen,
No skiffs were off the waters wild, for they'd never been,
The boats as usual, and their crews, right drunken crews, I ween,
Were moored around each coffee-house, *for hire,* were you green,
Of dimes you'd be delivered, in our transit up to town,

III.

Nor Alderman nor Commissaire were then upon the ground,
Amidst the swamped up Seventh rates they could not then be found,
They were boating with the "Leedies" for many miles around,
Or curing a mad dog's bite, with a hair from the same hound,
While ague and fever shook the frame of many a doleful wight.[27]

The anonymous writer "CROTCHET" pursued a similar theme in criticizing the lack of help extended by city government in the remainder of the poem that opens this chapter:

With astonishing munificence—they've provided for all,
Two or *three* skiffs—that will come at your call;
And by paying the boatman two dimes, or three,
Will quick take you home—wherever it be.

This, our town fathers, in their mirth and their glee,
Call giving each citizen their passage quite free.

But let me assure them, 'tis quite a mistake,
To think they will find each one such a "cake."

For some time past, a pious wish I have had;
(But the "Fathers," I fear, will think me too bad.)
'Tis, to tie them together, and do "nothing shorter"
Than to tumble them all into the dirtiest water.

There let them wollow [sic] and paddle about;
And no hand be stretched forth to help them get out,
Till they slink out themselves, dirty, nasty, and muddy;
And crawl home to their dens, and "our Engineer's" study.[28]

Repeatedly, anonymous contributors to the city's newspapers used satire to criticize the apparent lack of knowledge about how to close the crevasse. Captain Grant, Alderman Peters's handpicked favorite to fix the problem, failed to produce any positive results and abandoned the effort after ten days, on May 29. By that time, the crevasse was nearly three hundred feet wide. Some newspapers denounced Grant as an out-of-town interloper who was too old, a mere pile driver, and addicted to whiskey.[29] The Crevasse Committee met at the behest of the mayor on June 2 and considered dozens of submitted solutions for fixing the levee. An anonymous contributor to the *Weekly Delta* styling him- or herself "Nell" offered a despondent review of the committee's proceedings:

"Who made the Crevasse?" said the Mayor, with a sigh;
Up rose our great river, and said "It was I;
My bounds were too small, my waters confined,
And this to submit to I had not a mind.
With a sweep and a bound, I your levees did pass,
And that is the way that I made the Crevasse."

"Who said that he'd stop it—it should not go far?"
"Ah true, that was I!" whimpered Mr. Dunbar.
"But surely I mean, with a requisite force,
Which you did not give, as you know to your loss;

For force was to me like the strings to a poppet,
And that was the reason that I did not stop it."

"Who next tried to stop it!" again asked the Mayor;
"That, your Honor, was I, Capt. Grant, the surveyor;
And long did try, but no progress I made,
Yet, 'twas all for bad luck, your Honor, 'indade!'
So you'll surely not blame me, nor let the Crevasse
Be called by the name of 'Professor Grant's Pass.'"

"Who else tried the work?" Then arose a great lot
Of professors, surveyors, and I know not what;
And great was their noise, as they rushed to the chair
Of the Mayor, who presided, and great their despair,
As he waved them aside, and assumed the black cap,
And said, after giving the table a rap;

"My friends for your doom you impatiently wait;
Alas! It will come but too soon—not too late.
Therefore, hark to my words, and let it be known,
That 'ere sunset tomorrow, you all shall be thrown
Right into the Crevasse, which you've helped to make,
And then we shall see if you'll not close the break.
Above you this epitaph sad will I place,
Which Time nor Crevasse shall ever deface:
    'Within this Crevasse large do lie,
Men whom I thus condemned to die;
They promised all to stop this breach,
they would not practice—did but preach,
Our funds they wasted; so these men
Shall ne'er work our Crevasse again,
Unless, perchance, they wish to fix
Some Crevasse on the river Styx.
Peace to their shades! Oh! May their fates
A warning be to foolish pates,

Who, while with undue zeal they're warm,
Oft promise what they can't perform!'"[30]

With this Swiftian proposal to drown all of the city officials and engineers in the crevasse in order to stop it, "Nell" expressed the frustration that many New Orleanians felt as parts of the city fell underwater with no relief in sight. As Nell makes clear, at least some New Orleanians placed blame squarely on the various levels of city government and, by inference, demanded that these officials devise a solution quickly. Not only was the hole in the levee on Sauvé's plantation still open after a month, but citizens in the flooded areas saw little relief provided by government. The *Weekly Delta* produced a lengthy editorial entitled "The Crevasse Mania" in early June that also invoked satiric humor and classical allusions to make the point that politicians needed to stop interfering with local engineers and scientific principles and, instead, let the experts do what they were trained to do:

One of the most annoying, if not the most serious, discomforts, inflicted upon our people by the present inundation, is the prevalence among our citizens, of a most extraordinary mania, which leads many hitherto quiet and unobtrusive persons to imagine that they are Archimedeses, Michael Angeloes, Wrens, or Wattses—not as those eminent men of science, became such by long study and observations—but as Minerva leaped from the head of Jupiter, fully armed, to struggle with all the obstacles of nature. In the ecstasy of this mania, the old Latin proverb, *ne sutor ultra crepidam,* or every man to his trade, has been entirely lost sight of. The only persons who seem to have no definite plans for stopping the Crevasse are our professional engineers and surveyors. In general display of scientific knowledge, these gentlemen imitate the wisdom of Zeno, who remained silent, whilst the other babbling philosophers of Athens were displaying their resources to the Spartan ambassador, and being asked by the latter what message he had to send to the philosophers of Sparta, replied: "Tell them there is one man in Athens who had the wisdom to be silent."

But this fast and sudden development of scientific genius is due to the sagacity and liberality of wise and potent Signors who, in their well-cushioned chairs, concoct those mighty and profound ideas and plans, by which we are enabled to "live, move and have our being" in this patient city of ours. When an Alderman and Bank President [Samuel Peters] could so anathematize and confute a scientific man [George Dunbar], in his own peculiar sphere of knowledge and experience, as was done on a memorable occasion—and, alas! so fatally for our city—why should we wonder that every other grocer, cobbler, lawyer, schoolmaster, chemist and auctioneer considers himself equal to the emergency—the Moses, at the waving of whose wand the on-pressing waters shall gather up and recede?

... The most vulgar of all prejudices, is that which regards men of science and learning as unfit for great works, in which the principles of science are carried into practical operation. Some writer says that "your practical men are practical fools."... The man of science is the true practical man.... By science, we do not mean mere abstract rules, theories or systems contained in books, but the collected experience of men of research and study, who have devoted their faculties to the observation of the workings of nature.

... But we are happy to perceive that our authorities have recovered from the insanity, and are restored to common sense. They have determined, as we have all along contended they should do, to assign the labor of stopping the Crevasse to our regular engineers—the responsible officers of corporations, to whom are referred all the various plans, (*difficiles nugae*) of our Crevasse-crazy citizens. This is wise.

All now that is practicable or possible will be done. Let our Aldermen stay at home and paddle their own canoes, taking care to forward the necessary supplies to the engineers at the Crevasse.... The responsibility being thrown upon our engineers, they will not treat lightly any practical plan of operations. But our Aldermen—pleasant, agreeable, worthy citizens as they are, profound financiers, and stock-jobbers—let them hereafter remember the caution not to "play with edged tools." Let them abandon hydrostatical experiments, and save themselves from the opprobrious application of the last syllable of the disasters which they have striven to overcome, on "*practical principles*," by confining their appetite and wisdom to turtles and taxes.[31]

As "The Crevasse Mania" pointed out, after Grant's failure to stem the flow of the Mississippi through the crevasse on Sauvé's plantation, Dunbar and the surveyor from the First Municipality, Louis Surgi, were finally ordered by Mayor Crossman and the Crevasse Committee to return to the crevasse with as much money, men, and materials as they deemed necessary. Dunbar resumed work at Sauvé's plantation on June 3, and by June 20 the breach was finally closed. An aspiring poet who lived in the inundated Second Municipality's Seventh Ward penned a congratulatory rhyme, stanzas excerpted below, dedicated specifically to Dunbar while disparaging other city officials. Dunbar's star rose quickly as citizens even called for his election as the city's mayor. The poem also urged voters to refuse to reelect those politicians who had denied them aid during the flood.

I.
All hail thou Mississippi's mighty master,
Thou Macbeth of the Waters, hail—all hail!
Our feelings towards you are flowing faster
Than did the rush o'er which thou didst prevail.
Joy follows in the train of this disaster
Despite the sorrows which it must entail;
And I, for one would, as I am a sinner,
Go in for giving you a public dinner.

. . . .

VI.
Never, as I said, was there hope deferred,
Each day so long was sickening the heart,
And some one in the Council must have erred
To leave our Seventh boats and skiffs *to mart.*
For here throughout the neighborhood 'tis averred
They were a source of profit at the start.
    Small pleasure parties flitted o'er the deep,
    Some free as air of charge, some *not so cheap.*

VII.
I was a hydrophite, and had no boat;
My gunnels were all taken from my door,
And I had nothing whereupon to float;

Nothing to venture with—no post to moor
Even the smallest skiff to. Those who vote
For aldermen next year ought to be sure
    To elect a man who wouldn't in a miff
    Fall into hysterics at the term "skiff."

VIII.

'Tis not an un-euphonious term, either,
I've made it jingle very well in rhyme;
Yet some there are, no doubt, will find it "rather"
An indigestible concern. Some other time
The ward may turn on its conscript "Feyther,"
And lower him a pitch from the sublime.
    And send a harsher message to the ear
    Than that unholy sound it used to hear.

....

X.

Now, having stated my mind quite in full,
I bid farewell unto the inundation;
Perhaps, because my wits edge's getting dull,
For verse is not begot by procreation.
We cannot, spider-like, spin from the skull
Even to win a world's approbation.
    Yes, this small web I've wove, and here it is
    Gratis to both my friends and enemies.[32]

Outcomes

The hardship and deprivation that the inhabitants of New Orleans experienced in 1849 were genuine, but their humor and use of satire amid this disaster influenced public perceptions and helped bring real change to the city. Utilizing one of the few public tools available to them, New Orleanians penned creative and biting verse to comfort their fellow citizens and criticize their political officeholders. In response, the years immediately following the 1849 flood saw major changes in the way New Orleans's city government was organized and the way the government managed the levee system. New Orleans, along with adjacent Lafayette, was reorganized into a single municipality in 1852, giving the mayor's office more authority to handle emergencies. In the 1850 Swamp Land Act, which had been under consideration for several years, the federal government relinquished control of over 8 million acres in Louisiana to the state so that it could sell them to fund a well-engineered and unified levee system.

City politicians who had vociferously opposed the sufficient funding of flood prevention and resisted giving relief to survivors in 1849 faced intense criticism, with some of them, such as councilman Samuel Peters, resigning their office or declining to seek reelection. George Dunbar, on the other hand, was reelected easily as surveyor for the Second Municipality amid calls from some quarters that he instead run for mayor. Alas, the physical demands that Dunbar incurred in repairing Sauvé's levee, including nearly drowning, weakened him and brought on a recurrence of malaria that he had contracted years earlier, leading to his death in December 1850 at the age of thirty-eight. Other civil engineers gained new prestige and responsibilities, however, as the city, state, and national governments sought up-to-date scientific data and plans to prevent such catastrophes from happening again. Dozens of engineers flocked to New Orleans and offered a multitude of plans and reports, some fanciful and some prescient. Lastly, the decade of the 1850s brought renewed prosperity to New Orleans, fueled in part by city reform and engineering projects but also by the boom in cotton prices and other staple crop production shipped through New Orleans.

As the city resumed business and rebuilding in the summer of 1849 and optimism returned, satirical humor remained a way to discuss what had happened and why. Talk of a new railroad to link New Orleans via land with points north, "that will tap the Mississippi with more profit and less danger than Sauvé's crevasse did," became serious after the flood. A southern contributor to New York's *Spirit of the Times* enlightened readers: "By the way, did you ever hear the true origin of that affair? The fact is, the streets of Orleans—especially in the rear—had been filthy so long, that the Father of Waters was ashamed of the dirty son that had grown up by his side, and finding him too surly, or too stingy, to wash his own face, the Daddy undertook to do it for him. And he did it. But *Sauvé est sauve* [safe], as we say in French—and the Crevasse is closed. The streets for once are clean."[33]

As city employees, such as Dunbar's Surveyor's Office employees, dumped lime in the formerly flooded streets to clean up organic matter and prevent diseases in the days after the flood receded, nature helped the effort with a heavy rainfall that thoroughly cleaned the streets and, it was believed at the time, prevented any cholera outbreak for several months. In the years after 1849, responsibility for protecting New Orleans from flooding gradually shifted to the national government and the Army Corps of Engineers, but New Orleanians have continued to insist—often with biting satire—that their government, at whatever level, act aggressively to protect life and property.

NOTES

1. For the basic background and statistics, and the only scholarly piece devoted to the 1849 flood, see Harry Kmen, "New Orleans' Forty Days in '49," *Louisiana Historical Quarterly* 40, no.1 (1957): 25–45.

2. Scholars who have examined some aspects of the 1849 flood include: Ari Kelman, *A River and Its City: The Nature of Landscape in New Orleans* (Berkeley: University of California Press, 2003), 162–70; Craig E. Colten, *An Unnatural Metropolis: Wresting New Orleans from Nature* (Baton Rouge: Louisiana State University Press, 2005), 26–27; Donald W. Davis, "Historical Perspective on Crevasses, Levees, and the Mississippi River," in Craig E. Colten, ed., *Transforming New Orleans and Its Environs: Centuries of Change* (Pittsburgh: University of Pittsburgh Press, 2000), 84–106; and Ari Kelman, "Even Paranoids Have Enemies: Rumors of Levee Sabotage in New Orleans's Lower 9th Ward," *Journal of Urban History* 35, no. 5 (2009): 627–39.

3. Ted Steinberg, *Acts of God: The Unnatural History of Natural Disaster in America* (New York: Oxford University Press, 2000). For a global perspective on the human role in "natural" disasters, see Henrik Svensen, *The End Is Nigh: A History of Natural Disasters* (London: Reaktion Books, 2009).

4. See the column "Gross Imbecility" in the *Daily Delta* (New Orleans), May 12, 1849.

5. *Daily Crescent* (New Orleans), January 2 and 17, 1849.

6. *Weekly Delta* (New Orleans), March 5, 1849.

7. Ibid.

8. Ibid.

9. Ibid.

10. *Daily Crescent*, March 5, 1849.

11. Ibid., May 11, 1849; *Weekly Delta*, May 14 and 21, 1849.

12. *Daily Crescent*, May 14, 1849.

13. Ibid., May 17, 1849.

14. The same humorous column by "Cap'n Bender" appeared twice, in the *Daily Delta*, May 20, 1849, and the *Weekly Delta*, May 21, 1849.

15. *Spirit of the Times* (New York), June 30, 1849, 218.

16. *Daily Delta*, May 24, 1849.

17. D. Corcoran, *Pickings from the Portfolio of the Reporter of the New Orleans Picayune* (Philadelphia: Carey and Hart, 1846). See also: Norris Wilson Yates, *William T. Porter and the Spirit of the Times* (Baton Rouge: Louisiana State University Press, 1957), 12; Derek Colville, "History and Humor: The Tall Tale in New Orleans," *Louisiana Historical Quarterly* 39, no. 2 (1956): 161; M. Thomas Inge, ed., *The Frontier Humorists: Critical Views* (Hamden, Conn.: Archon Books, 1975), 3; James H. Justus, *Fetching the Old Southwest: Humorous Writing from Longstreet to Twain* (Columbia: University of Missouri Press, 2004), 242–43, 260–61; and M. Thomas Inge and Edward J. Piacentino, eds., *The Humor of the Old South* (Lexington: University Press of Kentucky, 2001), 95, 180–81.

18. Colville, "History and Humor," 166.

19. For studies of the role of satire in eighteenth- and nineteenth-century history, see: Bruce Ingham Granger, *Political Satire in the American Revolution, 1763–1783* (Ithaca, N.Y.: Cornell University Press, 1960); Elaine G. Breslaw, "Wit, Whimsy, and Politics: The Uses of Satire by the Tuesday Club of Annapolis, 1744 to 1756," *William and Mary Quarterly* 32, no. 2 (1975): 295–306; Alison G. Olson, "The Zenger Case Revisited: Satire, Sedition, and Political Debate in Eighteenth-Century America," *Early American Literature* 35, no. 3 (2000): 223–45; Oana Godeanu-Kenworthy, "The Political Other in Nineteenth-Century British North America: The Satire of Thomas Chandler Haliburton," *Early American Studies* 7, no. 1 (2000): 305–34; and Amy Wiese Forbes, *The Satiric Decade: Satire and the Rise of Republicanism in France, 1830–1840* (Lanham, Md.: Lexington Books, 2010).

20. *Daily Delta*, May 26, 1849.

21. *Mississippi Free Trader and Natchez Gazette*, May 30, 1849.

22. *Daily Delta*, May 27, 1849.

23. Ibid.

24. *Daily Crescent*, May 18, 1849.

25. *Weekly Delta*, May 21, 1849.

26. *Daily Delta*, May 27, 1849.

27. *Weekly Delta*, July 2, 1849.

28. "City Lyrios," *Daily Crescent*, May 28, 1849.

29. *Weekly Delta,* June 4, 1849.

30. Ibid., June 11, 1849; *Daily Delta,* June 7, 1849.

31. *Weekly Delta,* June 4, 1849.

32. Ibid., July 2, 1849, emphasis added to stanza 1.

33. *Spirit of the Times,* August 18, 1849, 26.

EPIDEMICS, EMPIRE, AND ERADICATION

Global Public Health and Yellow Fever Control in New Orleans

URMI ENGINEER WILLOUGHBY

By the end of the eighteenth century, yellow fever, a mosquito-borne viral infection, had become endemic in much of the Caribbean, Mexico, and Panama. In the nineteenth century, it appeared frequently in cities along the Gulf Coast of the United States, and constituted a major health hazard in New Orleans. It often had devastating effects, causing severe outbreaks in the 1850s and a disastrous epidemic that spread throughout the lower Mississippi Valley in 1878. Sporadic epidemics continued in the Gulf Coast region until 1905.

Despite the relative decline of acute epidemics between 1880 and 1905, federal interest in eradicating yellow fever solidified during this period, and, by 1905, public health authorities successfully eliminated the threat of yellow fever in New Orleans. Ideas about the danger of yellow fever among residents changed during this time, as common perceptions shifted from regarding the disease as a disruptive but tolerable public health problem, to emphasizing the disastrous potential of epidemics. For most of the nineteenth century, locals perceived the disease as an irritating yet inevitable hazard of the city's environment and resisted federal efforts to control its spread. By the end of the century, medical authorities believed that the threat of epidemic yellow fever constituted an issue of national, and even global, importance, and advocated public health measures to eradicate the disease. This ideological shift resulted from two interconnected factors: a belief in the ability of public health programs to eradicate yellow fever, and the growth of American imperialism in the Caribbean and Central America, which enabled the medical discoveries that generated faith in public health institutions.

Disease and Disaster Ecology

A complete and accurate understanding of why yellow fever was such a major problem in nineteenth-century New Orleans requires not only an awareness of the environmental conditions that allowed the spread of yellow fever, but also a consideration of the role of human endeavors in creating conditions that facilitated the rise of epidemics. The concept of the Anthropocene highlights the role of humans in altering the environment and the unintended consequences of human actions. As J. R. McNeill has articulated in his revision of Karl Marx's insights on the contingent nature of history, "people made their own history but they did not make it as they pleased because ecology would not let them."[1] In a process similar to the rise of yellow fever on plantations and cities in the Greater Caribbean during the seventeenth through nineteenth centuries, the disease emerged as a health hazard in antebellum New Orleans as a result of a series of ecological changes.

Yellow fever was a constant presence on the Gulf Coast from the start of the nineteenth century through the 1850s. Epidemics frequently occurred in New Orleans, Mobile, Galveston, Charleston, and cities on the coast of Florida. Ecological transformations that fostered the growth of yellow fever epidemics in New Orleans included both environmental and demographic changes. As in the larger Gulf Coast region and the Caribbean, *Aëdes aegypti* mosquitoes served as the primary insect host for the yellow fever virus. Environmental changes tied to the rise of sugar plantations and urban growth included extensive land clearance, swamp drainage, and construction, which expanded the habitat of *A. aegypti* by providing additional breeding sites and sources of nourishment. Canals, steamships, and railroads facilitated the spread of epidemics in the Mississippi Valley. The growth of trade and immigration in port cities further exacerbated the threat of epidemics by exposing the city to yellow fever every summer and providing the virus with a human population of nonimmunes.[2] Most healthy children who contract yellow fever experience only mild symptoms and survive, thus gaining lifelong immunity to the disease. Adults, on the other hand, are more likely to experience severe symptoms. Surviving the disease and acquiring immunity prevents a person from hosting the virus again. Therefore, only nonimmune individuals can spread the disease, and are in fact necessary for it to appear in epidemic form.[3] The disastrous potential of epidemics increased over the course of the nineteenth century as immigration to New Orleans increased. Further, when yellow fever spread to regions that had not previously encountered the virus, it decimated nonimmune populations.

From Miasmas to Bacteriology:
Medical Knowledge and Public Health

Throughout the nineteenth century, details regarding the etiology of yellow fever were unknown to physicians and public health officials in the United States. Despite the lack of detailed knowledge of transmission cycles and the role of mosquitoes, local physicians had a practical understanding of epidemiological patterns. Until the late nineteenth century, among medical professionals and residents of the Gulf Coast and Atlantic World, the dominant framework for understanding disease was broadly environmental; rather than searching for microscopic causes of disease, people understood illness as a result of ecological factors including climate, latitude, topography, soil conditions, temperature, changes in seasons, and meteorological events.[4] Throughout the antebellum era, medical practitioners in the Gulf Coast region drew connections between ecological patterns and diseases in warm climates, particularly fevers. Physicians often linked illness with the atmosphere, and theorized that diseases stemmed from local "foul airs" or "miasmas." Between 1848 and 1878, several researchers who practiced in the Gulf Coast and Caribbean region viewed yellow fever in terms of its ecological patterns, and connected the appearance of the disease to factors including urban growth and the presence of mosquitoes and/ or stagnant water, which served as a breeding ground for mosquitoes.[5] Yet these observations, though remarkably accurate, did not influence local government officials to pass any sort of public health legislation.

Environmental views of disease were connected to, and reaffirmed the ideas of, contemporary European colonial doctors who practiced in tropical regions. They generally considered diseases that they encountered in warm climates as distinct from familiar diseases that they found in temperate zones, and argued that European medical practices were insufficient in addressing tropical diseases.[6] Since New Orleans and southern Louisiana occupied a subtropical climate (from about 28 to 31 degrees north latitude), doctors viewed the city as an ambiguous space that existed physically and intellectually between the tropics and the temperate world. Changes in medical knowledge, stemming from the rise of bacteriology in the mid-nineteenth century, resulted in a rise in confidence that modern medicine could conquer diseases of the tropics.

Bacteriology, or germ theory, came to replace miasmatic theory in the late nineteenth century. It emerged in the late 1860s and rapidly became the mainstream ideology for understanding the causes of disease.[7] The bacteriological revolution was grounded in the pioneering laboratory work of researchers including French physician Louis Pasteur, German physician Robert Koch, British physician John Tyndall, Scottish physician Joseph Lister, and several others. Using microscopic examinations, test-tube cultures, and animal experiments, their investigations proved the existence of bacteria as a cause of disease. By the 1870s, bacteriology dominated disease-related research, and medical authorities throughout the world abandoned miasmatic ideology and searched instead for microscopic germs.[8] Indeed, from the 1870s through the 1890s, the bacteriological revolution catalyzed medical development throughout the world, influencing many researchers to study diseases by identifying microscopic organisms in the blood and organs of infected patients rather than observing more general environmental trends. Bacteriological theory dominated studies of yellow fever and public health initiatives aimed at yellow fever prevention. Because they could not see the yellow fever virus with contemporary microscopic technology, scientists were not able to identify the pathogen during this period. As a result of their focus on searching for a yellow fever germ, few medical researchers considered the theory of mosquito-borne transmission.

Medical views of diseases shaped popular attitudes about their danger and the efficacy of public health measures. In the mid-nineteenth century, responses to the lingering threat of severe yellow fever outbreaks in New Orleans ranged from blithe acceptance to severe panic among residents and visitors. Many inhabitants accepted the potential threat as a reality of life, such as A. Oakey Hall, who described the emptiness of the city during the summer, writing that "juleps and iced ale are in demand until the sunny hours of August, when Yellow Jack comes into town, and the room echoes to the tread of some score or so, whom death nor disease can frighten from the worship of the appetite; or who, secure by acclimation, over their clinking glasses or ice ringing goblets laugh at the passing terrors of the 'grim conqueror.'"[9] Fearful newcomers often fled the city during the summer. Core concerns among local elites included the economic and social consequences of both epidemics and preventative measures. State and local governments were hesitant to enforce public health regulations such as quarantine, which they feared would interfere with commerce, citing ineffectiveness as the primary reason.[10] Also, by the 1870s, bacteriology guided public health efforts, and the usual compromise was to promote cleansing and disinfection, which were ineffective in controlling yellow fever.

Public Health Debates Following the Epidemics of 1878–1879

Epidemics subsided during the Civil War and appeared to be less frequent and severe during Reconstruction, until an outbreak in 1878 initiated a regional epidemic in the Mississippi Valley, which spread to regions that had previously never been exposed to yellow fever. The epidemic was most pronounced in Memphis, Tennessee, and in New Orleans, where it caused nearly five thousand deaths, with a total of roughly twenty thousand deaths in the entire affected region.[11] It was the worst epidemic in New Orleans since 1853, both in human lives lost and in devastating economic impact. Moreover, many considered it the worst epidemic in the nation's history because it spread throughout the interior of the lower Mississippi Valley. It affected not only port cities along the Mississippi, but also smaller towns and plantations that had become connected to the river by railroads. The epidemic inspired a series of federal and local efforts to prevent future epidemics.

Immediately following the epidemic, Congress authorized a federal "Board of Experts" to investigate yellow fever and cholera in the southern states. Organized in Memphis, the board was composed of twelve medical authorities from various cities including Boston, Albany, Memphis, Savannah, Jackson (Mississippi), and Mobile. The board included three members from New Orleans: Dr. S. M. Bemiss of the University of Louisiana (now Tulane University); Dr. Stanford E. Chaillé, president of the New Orleans Board of Health; and Col. Thomas S. Hardee, a sanitary engineer. Many local medical authorities viewed the mandates of the federal board as intrusive to state-governed boards, which foremost considered the commercial interests of New Orleans. They criticized the board because it did not include more physicians from southern port cities, particularly Charleston and New Orleans, arguing that it functioned "under the leadership of the determined and managing apostle of quarantine."[12] Congress also gave "limited quarantine powers" to the Marine Hospital Service. Southern public health officials outside of New Orleans urged Congress to impose national quarantine regulations in New Orleans. Tennessee state representative Casey Young argued that "only a national quarantine, administered by a national board of health, could effectively exclude yellow fever."[13] In response to the possibility of federal quarantine in New Orleans, local medical authorities organized boards to address yellow fever prevention. These private organizations, including the New Orleans Parish Medical Society (1878) and New Orleans Auxiliary Sanitary Association (1879), promoted sanitation and "limited quarantine" as the best modes of preventing yellow fever and other diseases.[14]

By March of 1879, Congress had created the National Board of Health to assist with quarantines in southern ports. However, the board did not have the authority to effectively regulate local commercial activity. The board appointed twenty-five sanitary inspectors, with seventeen stationed in Memphis and New Orleans. The board instructed the inspectors "to act as an advisor or instructor, but never as possessing positive authority, unless such authority may be temporarily conferred to do so." The board maintained that it was "more important that [inspectors] exert all of the moral influence possible in order to effect the best results."[15] In the summer of 1879, yellow fever appeared again in both Memphis and New Orleans. While the disease caused nearly six hundred deaths in Memphis, less than twenty deaths were reported in New Orleans. The National Board of Health concluded that "through the efforts of the State board of health and the auxiliary sanitary association of New Orleans, assisted by the National Board, a very thorough cleansing and disinfection of the infected portion of the city was carried out. When this was effected the disease disappeared. How far this was a coincidence and how far a consequence, it is impossible as yet to assert."[16]

Despite unresolved disputes among the board members about the nature and prevention of yellow fever, they agreed on a list of public health recommendations. These recommended preventative measures, however, were not markedly different from previous regulations, which included the surveillance of ships, the strict enforcement of quarantine, disinfection of clothing and baggage, and the cleansing of bodies upon arrival. The board also emphasized isolation of sick persons, the segregation of the sick, and the establishment of "camps of refuge" outside of cities and towns where yellow fever had been epidemic in previous years. However, the board advised public health officials that "the apartments of the sick should . . . be freely ventilated," arguing that atmospheric air served as a disinfecting agent. In fact, the ventilation of these apartments may have actually contributed to the spread of yellow fever by allowing mosquitoes to move more freely between infected and uninfected persons. In general, the board made recommendations that did not reflect their field observations, which showed disinfection and sanitary reform were ineffective. For example, the board demonstrated the failure of chemical disinfectants in open air, but concluded that research and experimentation in disinfection practice "should be liberally encouraged."[17]

In April of 1879, the National Board of Health approved the establishment of the Havana Yellow-Fever Commission to conduct further research into the causes and prevention of yellow fever. Stanford Chaillé chaired the commission, and other members included George Sternberg, a U.S. Army physician and prominent bacteriologist who served as secretary; Juan (John) Guitéras, a Cuban pathologist; and Thomas Hardee as the board's sanitary engineer.[18] While in Cuba in 1879, Sternberg consulted with physician Carlos Finlay, but neither considered the role of the mosquito while working together. Both focused on investigating a bacteriological cause.[19] However, federal public health authorities ignored their research for more than twenty years, until the U.S. government took over the administration of Cuba after the Spanish-American War.[20]

The Spanish-American War and the War on Mosquitoes

U.S. involvement in the Spanish-American War altered the course of yellow fever research. By the spring of 1898, the United States had entered a military conflict that began as a Cuban revolt to gain independence from Spain in February of 1895. The Cuban-Spanish conflict threatened U.S. political and economic interests by disrupting trade and military control of the island, which was positioned as the "key" to control of the Gulf of Mexico.[21] The war also interfered with the possibility of the construction of an interoceanic canal through Panama, an idea that increasingly attracted entrepreneurs in the United States during the 1890s.[22] The United States defeated the Spanish in just three months, with fewer than four hundred war-related casualties. However, more than twenty-five hundred American soldiers died from yellow fever during the war, causing yellow fever control to become a priority among members of the U.S. Army Medical Board.[23] Between 1899 and 1900, military officers continued to enforce quarantine and sanitary reform, which included cleansing, disinfection, and fumigation.[24] The U.S. Army's occupation of Cuba served to promote yellow fever research because of the threat posed by the disease to officers stationed on the island. That threat prompted the federal government to adequately fund yellow fever research, which the board could carry out systematically in Havana.

Several institutional developments caused by the war eventually led to numerous scientific breakthroughs. Major organizational and educational reforms included the formation of the Army Nurse Corps, the Department of Military Hygiene at the U.S. Military Academy, and the Army Medical Reserve Boards. The Marine Hospital Service established a number of boards designed to target tropical diseases, focused on typhoid fever and yellow fever in the Americas. In May of 1900, Sternberg, who had begun serving as surgeon general of the Marine Hospital Service in 1893, appointed a board to investigate yellow fever in Cuba, known as the U.S. Yellow Fever Commission or Reed Board, directed by Walter Reed of Virginia, who was working as a professor of bacteriology and clinical microscopy at the Army Medical School at the time. The board's three other members were James Carroll, a Canadian bacteriologist; Jesse Lazear of Baltimore, a bacteriologist who had begun studying malarial parasites in the 1890s; and Cuban physician Aristides Agramonte, who oversaw autopsies and pathological research.[25]

Initially, the Reed Board attempted sanitary enforcement. They did not go to Havana with the intent of eradicating mosquitoes. Instead, the board researched the cause of yellow fever while advising public health officers to enforce sanitary policy, particularly cleansing and disinfection.[26] They also conducted a series of bacteriological investigations, including a lengthy study disproving Italian bacteriologist Giuseppe Sanarelli's theory that yellow fever was caused by a bacterium that he identified as *Bacillus icteroides*.[27] Reed appointed U.S. Army medical officer and yellow fever immune W. C. Gorgas of Mobile, who had been stationed in Cuba since the beginning of the war, as chief sanitary officer in Havana. Gorgas wrote that, when they arrived in Havana in 1898, "the military authorities concluded that this was the opportunity which the United States has been awaiting for the past two hundred years. Thinking that yellow fever was a filth disease, they believed that if we could get Havana clean enough, we could free it from yellow fever. It was felt that if we could eliminate Havana as a focus of infection, the United States would cease to be subject to epidemics."[28] By late summer, Gorgas believed that "Havana had been cleaner than any other city had ever been up to that time." However, he noted that "yellow fever had been steadily growing worse than since we had taken possession of the city, and in 1900 there were a greater number of cases than there had been for several years."[29]

Due to their collaboration with Cuban researchers, the commission began testing a theory put forth by Carlos Finlay in the late 1870s, which claimed that *Culex* mosquitoes served as a host of yellow fever. While most medical researchers continued to search for a microscopic yellow fever–causing germ, Jesse Lazear began conducting mosquito experiments, using eggs that he had obtained from Finlay. After infecting himself and James Carroll with yellow fever, Lazear died in September of 1900 while Reed was in Washington D.C.[30] After Lazear's death, Reed became convinced that the mosquito was the yellow fever vector, and decided not to inoculate himself when he returned to Cuba. By November, Reed established Camp Lazear in Quemados, an isolated area southwest of Havana. His team included a number of Cuban physicians, such as Finlay, Juan Guitéras, and A. Díaz Albertini.[31] The researchers at Camp Lazear systematically tested Finlay's mosquito-transmission theory on paid volunteers, consisting of soldiers and newly arrived Spanish immigrants.[32] After confirming the mosquito theory, and determining the exact incubation time, Reed sent samples of the species to Washington, where L. O. Howard, an entomologist employed by the U.S. Department of Agriculture, specified them as *Culex fasciatus*.[33] The decision to pursue the hypothesis of mosquito-borne yellow fever transmission was a result of American access to research and testing in the Cuban environment, as well as the influence of transnational intellectual currents, as British imperial physician Patrick Manson's research on the role of *Anopheles* mosquitoes in transmitting malaria gained acceptance.

The enforcement of yellow fever eradication campaigns, which targeted mosquitoes, ultimately led to the end of yellow fever in Havana. The experiments at Camp Lazear, which lasted only a few months, resulted in the funding and enforcement of mosquito-suppression efforts in Havana, led by Gorgas, who became known for eradicating yellow fever in the city for the first time in nearly two hundred years.[34] Gorgas and his team engaged in operations aimed to both kill *A. aegypti* and prevent it from breeding and biting humans. They used pyrethrum powder to kill mosquitoes, alongside a number of other anti-mosquito measures including oiling still-water receptacles, draining lowlands, and screening buildings that housed infected humans, including hospitals and private homes.[35]

The triumph of public health measures in Cuba resulted from the enforcement of mosquito-eradication policies, based on the acceptance of Finlay's mosquito-transmission theory among U.S. public health authorities, including Sternberg and the Reed Board. The eradication of yellow fever in Cuba at the turn of the century can most accurately be described as a collaboration between Finlay and a group of North American and Cuban bacteriologists, fiscally and militarily supported by the U.S. government. This success led to the emergence of global mosquito-eradication campaigns, led by the United States.

Imperialism and Eradication Campaigns

In 1902 the U.S. Public Health Service expanded and merged with the Marine Hospital Service, forming an organization known as the U.S. Public Health and Marine Hospital Service. Essentially, the federal bureau acquired a military component.[36] When Sternberg died in 1902, Walter Wyman, who had been surgeon general of the Public Health Service since 1891, took over his duties and became head of the organization. This merger enabled the service to enforce public health measures within the nation and overseas, and facilitated military campaigns during this period of U.S. imperial expansion into tropical and subtropical regions in Cuba, Puerto Rico, Mexico, the Panama Canal Zone, Guam, Hawaii, and the Philippines.[37]

At this time, European medical researchers from England, France, and Germany began establishing schools of "tropical medicine," including the Liverpool School of Tropical Medicine (1898) and the London School of Tropical Medicine (1899), while a number of specialists conducted research in Brazil. When the Reed Board began investigating the role of mosquitoes in the propagation of yellow fever in Havana in 1900, the Liverpool School of Tropical Medicine appointed several doctors to study the disease in Para. By the end of the following year, a French coalition began an expedition to study the transmission of yellow fever in Brazil and established its headquarters in Rio de Janeiro, and by 1904 the Hamburg School of Tropical Medicine sent doctors to study the disease in Brazil, also based in Rio. By 1905, the Liverpool School of Tropical Medicine had reestablished its Yellow Fever Laboratory in Para under the direction of new physicians.[38]

Mosquito-Eradication Campaigns in the Atlantic World

Between 1900 and 1909, as news of the success of the Reed Board in Cuba traveled throughout the Caribbean and tropical Atlantic, public health authorities initiated several anti-mosquito campaigns. These typically included a military component, which enabled the systematic enforcement of anti-mosquito measures by public health officials. A national yellow fever board enforced mosquito eradication in Brazil, while U.S. and British boards operated in Mexico, Puerto Rico, Panama, Belize, Honduras, numerous island colonies throughout the Caribbean, and the southern United States.

In Brazil, public health authorities successfully eradicated the disease for several years by engaging in similar practices as the Reed Board in Havana. A group of researchers in Rio de Janeiro including Domingo Pereira Vaz, Oscar Marques Moreira, Januario Fiori, André Ramos, and Emilio Ribas tested the mosquito-transmission theory by allowing infected mosquitoes to bite them. By March of 1903, after his appointment as director of Brazil's federal Department of Public Health, the "young, virtually unknown" Oswaldo Cruz sent a representative to Havana to observe the work of American authorities in Cuba, and by April organized the new Serviço de Prophylaxia da Febre Amarella (Yellow Fever Prophylaxis Service).[39] The service operated by dividing Rio de Janeiro into sanitary districts, policing suspected areas, destroying mosquitoes, and identifying and isolating yellow fever patients. They formed "anti-mosquito brigades," one of which "consisted of 1,500 men to wage relentless war upon all the breeding places of the *Stegomyia*." In addition to destroying breeding spaces, brigades attempted to kill mosquitoes using sulfur and pyrethrum.[40] By the end of the year, "the brigades of 'mosquito killers' became a familiar sight in the city."[41] Incidences of yellow fever decreased from 1903 until the disease was absent from the city in 1909. Mosquito brigades also successfully removed the threat of yellow fever in Santos, in addition to island colonies throughout the Caribbean.[42]

The U.S. Marine Hospital Service began sending officials to Mexico and Puerto Rico to conduct yellow fever research in 1900, focusing on the study of *Stegomyia* mosquitoes. Between 1900 and 1903, laboratories in Washington analyzed specimens collected by public health officers stationed in Laredo (Texas), Ponce (Puerto Rico), Tampico, and Vera Cruz (Mexico).[43] Many officers who were stationed in these areas conducted research, rather than overseeing eradication campaigns. Officers who had acquired immunity to yellow fever could safely work in these regions, collecting mosquitoes and observing their breeding and biting patterns. Scientists in Washington who received the mosquitoes could run tests in a controlled environment, allowing them to find effective methods of eliminating breeding spaces and killing the mosquitoes, in addition to developing technologies to protect humans while traveling or sleeping. By 1904, the U.S. Public Health and Marine Hospital Service advised public health officers to burn pots of sulfur in order to kill mosquitoes, but determined that "the disinfection of baggage and passengers' effects to prevent infection from yellow fever is no longer required."[44] Officials redefined and systematically enforced quarantine regulations, which focused on destroying mosquitoes and their breeding spaces and removing sick passengers, rather than detaining and cleansing ships.

The success of the Reed Board enabled U.S. representatives to seriously consider the construction of an interoceanic canal through Panama. However, the prospect raised concerns about yellow fever epidemics, which commonly occurred in the Canal Zone that was bordered by Colón on the Atlantic and Panama City on the Pacific. French engineers had repeatedly attempted to build a canal through Panama since 1880, but their efforts were unsuccessful in large part because of the presence of yellow fever. Impediments to the French project included labor shortages because of yellow fever deaths, in addition to environmental obstacles related to the rocky landscape, tropical storms, and malaria.[45]

After the Spanish-American War, federal officials and local entrepreneurs in the United States became increasingly interested in the prospect of the canal, which would give traders in the Gulf of Mexico direct and rapid access to goods from Asia. After the United States pledged to build the canal in 1902, Patrick Manson, who served as medical adviser to the British Colonial Office at the time, warned that, without the strict enforcement of anti-mosquito measures, yellow fever could cause devastating epidemics in Asia. He emphasized the disruption of trade that yellow fever would cause if it "broke out in distributing centres such as Zanzibar, Aden, Bombay, Calcutta, Colombo, Singapore, Batavia, Bangkok, Saigon, Hong Kong, Shanghai, or Yokohama."[46] He concluded that quarantine was unnecessary because the journey from Panama to Asia took more than two weeks, and in this time sick individuals would no longer be infectious.[47] Despite his concerns, Manson was optimistic about the ability of U.S. officials to contain the disease. He applauded the success of the Reed Board, and advised public health officers in the Canal Zone "to stringently enforce destruction of all mosquitoes, their eggs, and their larvae in all ships, before these ships are allowed to clear from the Pacific end of the Canal, or from other ports on the Pacific side of the American Yellow Fever District."[48] In the fall of 1902, Sternberg relieved Gorgas of his duty in Havana and placed him in charge of sanitary work on the Isthmus of Panama. During the year prior to the commencement of the building of the Panama Canal, Gorgas went to Egypt as representative of the U.S. Army to the first Egyptian Medical Congress, where he was directed to "examine into what had been the sanitary conditions during the construction of the Suez Canal" more than thirty years earlier.[49]

The eradication of yellow fever in the Canal Zone, directed by Gorgas beginning in 1903, elicited support for the construction of the canal among federal representatives and local commercial and public health boards. The prospect of a canal especially drew the support of the Board of Trade in New Orleans, which declared in 1904 that Panama "holds the key to the commercial situation of the great nations, and when our Government shall have completed the gigantic obligation it has assumed, and the canal is open for traffic, there will be thrust upon us an opportunity which, if not taken advantage of to the maximum, will mean our commercial undoing."[50] The board asserted that the Republic of Panama "is today the centre of the world's greatest activity," and considered New Orleans to be "the gateway to the South and the Mississippi Valley at this period of the world's progress."[51] Members of the Board of Trade emphasized the importance of yellow fever control and affirmed the success of the Reed Board. They reported that "Dr. Gorgas has had wonderful experience with health matters in Havana, and was instrumental in eliminating yellow fever in Cuba, and on the Zone, where the regulations, health conditions, and the control of the sanitation, health laws, and hospital service, are absolutely under his control, he expects to be equally successful," and petitioned the Louisiana State Board of Health to cooperate with the U.S. Public Health and Marine Hospital Service.[52] Between 1903 and 1908, Gorgas enforced policies aimed to protect citizens from mosquitoes and prevent breeding, such as prohibiting the collection of stagnant water by screening cisterns, enforced by house-to-house inspections and fines for those who failed to comply with public health mandates.[53] Construction began in 1904, one year after Gorgas's mosquito-eradication campaign successfully removed the threat of yellow fever in the Canal Zone.

Campaigns in Cuba, Brazil, Mexico, Puerto Rico, and Panama paved the way for the acceptance and enforcement of campaigns in the United States. In 1905, rumors of the presence of yellow fever in the southern states prompted the Public Health and Marine Hospital Service to send officers to investigate in cities throughout Louisiana, Mississippi, Alabama, Georgia, Texas, and Florida.[54] In addition to carrying out government research, public health officers were ordered "to conduct a campaign of education among the medical profession and laity . . . upon the importance of screening all cases of febrile diseases from the access of mosquitoes until a positive diagnosis is made, and upon methods for the destruction and propagation of these insects."[55] For the first time, federal public health officers led a national campaign against mosquitoes on the Gulf Coast of the United States.

Yellow Fever Eradication in New Orleans

The last yellow fever epidemic in the United States occurred in New Orleans the summer of 1905. On July 21, public health authorities announced that yellow fever was officially present in the city. Representatives from the city and state boards of health, health officers from surrounding states, as well as the Public Health and Marine Hospital Service, agreed to conduct mosquito-prevention campaigns in addition to quarantining the port of New Orleans.[56] The success of the 1905 campaigns can be attributed to the cooperation of federal, state, and local organizations, including the Public Health and Marine Hospital Service, the Orleans Parish Medical Society, the Citizens' Volunteer Ward Organization, and several other local media and educational efforts.[57] The systematic enforcement of mosquito-eradication policies, directed by Public Health and Marine Hospital Service officers, effectively removed the threat of yellow fever from the city by engaging in "general warfare against all mosquitoes, except swamp."[58] The service employed more than twelve hundred men and placed each ward under the supervision of an army surgeon, instructed to discover and isolate early cases of yellow fever, kill all *Stegomyias,* and ensure that each ward was "equipped with its forces of inspectors, oilers, screeners [and] fumigators." Medical officers used practices similar to those employed in Cuba, Brazil, and Panama, including the screening and oiling of cisterns, and fumigation using sulfur, pyrethrum, and steam.[59] The screening and oiling of cisterns in particular proved effective in decreasing the proliferation of mosquitoes.

By this time, local ideas about yellow fever control had shifted. Many residents actively supported the anti-mosquito policies. Young proponents wore buttons that read, "My cisterns are all right; How are yours?" Every week, officials promoted a "mosquito killing day," on which they encouraged all residents of the city to kill the mosquitoes in their homes by fumigation. With the cooperation of local organizations and individuals, the Public Health and Marine Hospital Service successfully eliminated the threat of yellow fever in New Orleans.[60]

The success in New Orleans was followed by more collaborations between the U.S. Public Health and Marine Hospital Service and the Liverpool School in Central America and the British Caribbean. In 1905, members of the Liverpool School worked with the U.S. Public Health and Marine Hospital Service to conduct eradication campaigns in Central America, British and Spanish Honduras, Belize, and adjacent republics, at ports that supplied fruit to the United States, often passing through New Orleans.[61] Within a few years, they began a number of yellow fever operations in British Caribbean colonial possessions, including Barbados, St. Lucia, Demerara, Trinidad, Grenada, and Jamaica.[62] British colonial officials worked with local authorities to carry out education campaigns, in addition to a "war against insect pests in the West Indies," which consisted of "a progressive policy of extermination waged against the *Stegomyia*," and a number of antilarval measures aimed at eliminating open stagnant water containers.[63] British public health officials in Caribbean colonies advocated a policy that was based on the success of the U.S. Yellow Fever Commission. British pathologist Rubert Boyce, who assisted in the public health expeditions to British Honduras and New Orleans and was a founding member of the Liverpool School of Tropical Medicine, declared that "a policy of this kind is bound to bear fruit, as it has already done in New Orleans, Cuba, Panama, Brazil, Mexico, etc. etc., and I am convinced, bring about the total eradication of yellow fever."[64] In 1910, Boyce remarked, "yellow fever is not to-day regarded as the inevitable penalty of our desire to go to tropical lands; it is to-day the penalty of ignorance and superstition."[65] By 1912, this global movement initiated the establishment of the nation's first School of Hygiene and Tropical Medicine in New Orleans.

The Rise of Tropical Medicine and the Military-Medical Complex

The development of the field of tropical medicine was intertwined with the growth of what I term a "military-medical complex," an institutional pattern in which collaboration with military officers enabled public health enforcement in global, imperial spaces. Medical officers have long been part of military operations, with the purpose of providing medical care to soldiers. However, this institutional development marks a substantive change in the operations of medical officers, as they became an essential component of medical research and public health enforcement.

Medical researchers studied diseases of warm climates throughout the colonial era, before the formal discipline of tropical medicine emerged in the late nineteenth century. Studies of diseases that were particular to "warm climates" preceded the bacteriological era, and in fact merged with miasmatic theory. Both the study of medicine of warm climates and tropical medicine were based on the idea that the tropics were inherently different from the temperate world, not only in geography and climate, but also in culture, race, and social patterns. Many doctors who studied diseases of warm climates or tropical medicine also studied race, and in most cases, their racial preconceptions are evident in their research.

Although the field of tropical medicine is often associated with the colonial world, parts of the southern United States, including New Orleans, fit into the pattern of tropical medical study. Subtropical areas like New Orleans inhabited a place between the tropical and temperate zones, and contemporaries viewed the city as a liminal space in terms of race, culture, and society as well.[66]

The field of tropical medicine that developed in European medical institutions between 1890 and 1910 reflected a long history of colonial research directed towards diseases of warm climates. Many historians consider British colonial physician Patrick Manson, who researched the role of parasitic organisms and mosquitoes in malaria and founded the London School of Tropical Medicine in 1899, as the founder or "father" of tropical medicine.[67] During the late 1890s, while working in Xiamen (Amoy), Manson researched the role of the *anopheles* mosquito in the transmission of malaria. Also at this time, Ronald Ross, an officer in the British Indian Medical Service stationed outside of Hyderabad, studied the *plasmodia* parasites that cause malaria.[68] Just as Manson and Ross conducted research in British imperial spaces, the team of U.S. Army surgeons credited with the discovery of the etiology of yellow fever conducted experiments in Cuba between 1900 and

1902, when the United States occupied the island after the Spanish-American War.

In his study of Manson and the "conquest of tropical disease" in the British Empire, Douglas M. Haynes argues that imperialism was the driving force behind the rise of tropical medicine. He contends that "the history of tropical medicine is more than a story about disease in the tropical world. It reveals the critical role of imperialism in constituting British medicine and science in the nineteenth and twentieth centuries. The health care needs of the formal and informal British Empire contributed to the growth, as well as the institutional development, of the profession at home."[69] Haynes argues that malaria "posed the single greatest challenge to the expansion of European colonies."[70] Similarly, yellow fever presented the greatest threat to the expansion of the United States in the Caribbean and Panama, despite the presence of malaria in the region. The health care needs of American soldiers stationed in Cuba during the U.S. occupation were a critical factor in the establishment of medical institutions and organizations aimed at eliminating yellow fever.

The growth of a national military-medical complex, in which the federal government relied on military support to implement and enforce sanitary measures, was integral to the successful eradication of yellow fever in locales throughout the tropical Americas. Having military support played an essential role in the implementation of new sanitary measures. Only a large, intrusive, and coercive public health force could effectively fight a disease like yellow fever. After the triumph of the U.S. Army Board in Cuba and Panama, military involvement and the establishment of military institutions grew and continued to practice sanitary policies in territories outside of the nation. Federal imperial interest in expansion into the tropical Caribbean led to the growth and development of public health and medical research institutions in the United States, including the establishment of the U.S. Public Health and Marine Hospital Service (1902), the American Society of Tropical Medicine (1904), and the Walter Reed Army Medical Center (1909).

The institutionalization of medical services in military regimes is a global pattern. Scholarship in the field of tropical medicine in the nineteenth century has shown a consistent pattern in the emergence and growth of state medical policies in the colonial and imperial regimes of the late nineteenth century, particularly in the British Empire. Historian Julyan Peard suggests that, in addition to historical studies of tropical medicine in colonial and imperial contexts, the history of tropical medicine in South America belongs in this historiography. He argues that, "whether a country traced its medical heritage back to the colonial era or to the national era, it was certainly the case that in the second half of the nineteenth century all of the largest Latin American nations trained their doctors locally and followed a Western European tradition of medical education."[71] In Brazil, the Bahian Tropicalista School of Medicine, active between 1865 and 1890, consisted of a group of Brazilian and foreign physicians who studied diseases that were found in Brazil, who "made new discoveries in parasitology, contributed to ongoing debates on beriberi, parasitology, leprosy, tuberculosis, dracontiasis, and helped reformulate the accepted pattern of Brazilian nosology."[72]

The history of the eradication of yellow fever in New Orleans at the turn of the century reveals links between imperialism and the rise of tropical medicine. Medical knowledge regarding the role of the mosquito explained the mysterious patterns of yellow fever and the ineffectiveness of previous sanitary campaigns, and led public health officials to focus on a new enemy: the mosquito. Attempts to study yellow fever in tropical regions continued in the twentieth century and remained linked to colonial and imperial regimes, most notably in West Africa. Despite global efforts to control the disease, cases of yellow fever have increased in equatorial Africa and South America.[73] Further studies on the ecological consequences of public health efforts directed towards the destruction of mosquitoes are likely to reveal unintended impacts on both mosquito and human populations.[74]

NOTES

1. From an ecological perspective, McNeill reiterates Marx's argument that "men make their own history, but they do not make it as they please" (J. R. McNeill, *Mosquito*

Empires: Ecology and War in the Greater Caribbean, 1620–1914 [Cambridge, U.K.: Cambridge University Press, 2010], 3–4). See also Will Steffen, Paul J. Crutzen, and J. R. McNeill, "The Anthropocene: Are Humans Now Overwhelming the Great Forces of Nature?" *Ambio* 36, no. 8 (December 2007): 614.

2. Entomologists have described *A. aegypti* as an "urban species" that thrives in built environments, with a unique preference for breeding in man-made containers. See: James D. Goodyear, "The Sugar Connection: A New Perspective on the History of Yellow Fever," *Bulletin of the History of Medicine* 52 (1978): 12; S. R. Christophers, *Aëdes Aegypti (L.), the Yellow Fever Mosquito: Its Life History, Bionomics and Structure* (Cambridge, U.K.: Cambridge University Press, 1960), 57; McNeill, *Mosquito Empires,* 47–50; Charles Morrow Wilson, *Ambassadors in White: The Story of American Tropical Medicine* (New York: Henry Holt and Co., 1942), 269. See also: Urmi Engineer, "Sugar Revisited: Sweetness and the Environment in the Early Modern World," in *The Global Lives of Things: The Material Culture of Connections in the Early Modern World,* ed. Anne Gerritsen and Giorgio Riello (New York: Routledge, 2016): 198–220, and Urmi Engineer Willoughby, *Yellow Fever, Race, and Ecology in Nineteenth-Century New Orleans* (Baton Rouge: Louisiana State University Press, 2017).

3. Sareen E. Galbraith and Alan D. T. Barrett, "Yellow Fever," in *Vaccines for Biodefense and Emerging and Neglected Diseases* (Amsterdam: Elsevier, Inc., 2009), 756; Philip D. Curtin, "'The White Man's Grave': Image and Reality, 1780–1850," *Journal of British Studies* 1, no. 1 (1961): 96.

4. Nancy Tomes, *The Gospel of Germs: Men, Women, and the Microbe in American Life* (Cambridge, Mass.: Harvard University Press, 1998), 32–33; Nancy Stepan, *Picturing Tropical Nature* (Ithaca, N.Y.: Cornell University Press, 2001), 153.

5. E. D. Fenner, "Reports from Louisiana: Article I," *Southern Medical Reports* 1 (1849): 23; E. H. Barton, "Report upon the Meteorology, Vital Statistics and Hygiene of the State of Louisiana," *Southern Medical Reports* 2 (1850): 135–37; *Report of the Sanitary Commission of New Orleans on the Epidemic Yellow Fever of 1853, Published by Authority of the City Council of New Orleans* (New Orleans: Picayune Office, 1854), 389–90. See also Craig E. Colten, *Southern Waters: The Limits to Abundance* (Baton Rouge: Louisiana State University Press, 2014), 92–93.

6. Mark Harrison, "'The Tender Frame of Man': Disease, Climate, and Racial Difference in India and the West Indies, 1760–1860," *Bulletin of the History of Medicine* 70, no. 1 (1996): 71–72.

7. Here I am considering the broad consequences of bacteriology as an ideology or cultural system. Philip A. Howard has articulated a useful definition of the term *ideology* that goes beyond its "traditional construction as the content of a set of beliefs and ideas," and includes "the social processes, institutions, and arrangements manufactured by a society's culture." See Howard, *Black Labor, White Sugar: Caribbean Braceros and Their Struggle for Power in the Cuban Sugar Industry* (Baton Rouge: Louisiana State University Press, 2015), 239; Christopher Hamlin, "Bacteriology as a Cultural System: Analysis and Its Discontents," *History of Science* 49 (September 2011): 269–70.

8. Tomes, *Gospel of Germs*, 32–33.

9. A. Oakey Hall, *The Manhattaner in New Orleans; or, Phases of "Cresent City" Life* (Baton Rouge: Louisiana State University Press, 1976 [1851]), 10–11.

10. C. B. White, *Disinfection in Yellow Fever as Practised at New Orleans in the Years 1870 to 1875 Inclusive* (Washington, D.C.: American Public Health Association, 1876), 7, 14.

11. Jo Ann Carrigan, *The Saffron Scourge: A History of Yellow Fever in Louisiana, 1796–1905* (Lafayette: Center for Louisiana Studies at the University of Southwestern Louisiana, 1994), 127, 404.

12. "Editorial," *Atlanta Medical and Surgical Journal* 16, no. 10 (January 1879): 625.

13. *Memphis Daily Avalanche*, March 11, 1879, qtd. in Margaret Humphreys, *Yellow Fever and the South* (Baltimore: Johns Hopkins University Press, 1999), 63–64.

14. Auxiliary Sanitary Association of New Orleans, *An Address from the Auxiliary Sanitary Association of New Orleans to the Other Cities and Towns in the Mississippi Valley* (New Orleans: L. Graham Book Printer, 1879), 8; John H. Ellis, *Yellow Fever and Public Health in the New South* (Louisville: University of Kentucky, 1992), 86.

15. National Board of Health, *Annual Report* (Washington, D.C.: Government Printing Office, 1879), 13.

16. Ibid., 19. Although only 19 deaths were reported in New Orleans, more than 150 were reported in Louisiana.

17. *Conclusions of the Board of Experts Authorized by Congress to Investigate the Yellow Fever Epidemic of 1878...* (Washington, D.C.: Judd & Detweiler Printers, 1879), 26–27.

18. National Board of Health, *Annual Report*, 33.

19. Nancy Stepan, "The Interplay between Socio-Economic Factors and Medical Science: Yellow Fever Research, Cuba and the United States," *Social Studies of Science* 8, no. 4 (November 1978): 403; Carlos Eduardo Finlay, Carlos Juan Finlay, and Morton Charles Kahn, *Carlos Finlay and Yellow Fever* (Oxford, U.K.: Institute of Tropical Medicine of the University of Havana by Oxford University Press, 1940), 58–59.

20. Stepan, "Interplay between Socio-Economic Factors and Medical Science," 403; Rudolph Matas, Frederick William Parham, and Thomas Smith Dabney, *Special Articles on Yellow Fever* (New Orleans: L. Graham & Son, Ltd., 1897), 12.

21. Robert T. Hill, *Cuba and Porto Rico with the other Islands of the West Indies: Their Topography, Climate, Flora, Products, Industries, Cities, People, Political Conditions, etc.* (New York: Century Co., 1899), 33; José Cantón Navarro, *History of Cuba: The Challenge of the Yoke and the Star* (Havana: SI-MAR S.A., 2001), 9.

22. Vincent J. Cirillo, *Bullets and Bacilli: The Spanish-American War and Military* (New Brunswick, N.J.: Rutgers University Press, 2004), 6.

23. Ibid., 1.

24. Stepan, "Interplay between Socio-Economic Factors and Medical Science," 409.

25. Wilson, *Ambassadors in White*, 102–3.

26. William Crawford Gorgas, *Results Obtained in Havana from the Destruction of the Stegomyia Fasciata Infected by Yellow Fever* (Havana: Sanitary Department, 1902), 11.

27. William Crawford Gorgas, *Sanitation in Panama* (New York: Appleton & Co.), 7–8; E. K. Sprague, "Present Status of the Bacteriology of Yellow Fever," in *Yellow Fever: Its Nature, Diagnosis, Treatment, and Prophylaxis, and Quarantine Regulations Relating Thereto* . . . (Washington, D.C.: Government Printing Office, 1898), 172–73; Just Touatre, *Yellow Fever: Clinical Notes,* trans. Charles Chassaignac (New Orleans: New Orleans Medical and Surgical Journal Limited, 1898), 153–54.

28. Gorgas, *Sanitation in Panama,* 5.

29. Ibid., 6.

30. James Carroll died in 1907 of myocarditis, which his physicians attributed to the severe attack of yellow fever that he contracted in 1900. Several newspapers asserted that Walter Reed also died as a result of his yellow fever experiments, but in fact he died in 1902 from appendicitis. See Martha Sternberg, *George Miller Sternberg: A Biography* (Chicago: American Medical Association, 1920), 277.

31. *Annual Reports of the War Department for the Fiscal Year ended June 30, 1901; Reports of Chiefs of Bureaus* (Washington, D.C.: Government Printing Office, 1901), 727.

32. Ibid., 726; Michael McCarthy, "A Century of U.S. Army Yellow Fever Research," *Lancet* 357 (June 2001): 1772.

33. Entomologists have since reclassified this mosquito numerous times. By 1901, medical authorities in the United States had renamed the species *Stegomyia fasciata* (*S. fasciata*). Until the 1950s, medical researchers classified the mosquito now known as *A. aegypti* as the *S. fasciata* or *S. calopus* (before 1930). See Gorgas, *Sanitation in Panama,* 16–17; Aristides Agramonte, "The Inside History of a Great Medical Discovery," *Scientific Monthly,* December 1915, 219; Leland O. Howard, Harrison G. Dyar, and Frederick Knab, *The Mosquitoes of North and Central America and the West Indies* (Washington, D.C.: Carnegie Institute, 1912), 293–94; George Augustin, *History of Yellow Fever* (New Orleans: Searcy & Pfaff Ltd., 1909).

34. Cirillo, *Bullets and Bacilli,* 118.

35. Gorgas, *Results Obtained in Havana from the Destruction of the Stegomyia Fasciata,* 12–13.

36. In 1912, the U.S. Public Health and Marine Hospital Service was renamed the Public Health Service.

37. Rubert Boyce, *Mosquito or Man? The Conquest of the Tropical World* (London: John Murray, 1910), 157.

38. Rubert Boyce, *Yellow Fever Prophylaxis in New Orleans, 1905* (London: Williams & Norgate, 1906), 6–7.

39. Nancy Stepan, *Beginnings of Brazilian Science: Oswaldo Cruz, Medical Research and Policy, 1890–1920* (New York: Science History Publications, 1976), 88–89.

40. Boyce, *Mosquito or Man?,* 188.

41. Stepan, *Beginnings of Brazilian Science,* 88–89.

42. Ibid., 91; Rubert Boyce, *Yellow Fever and Its Prevention: A Manual for Medical Students and Practitioners* (London: John Murray, 1911), 295–96, and *Mosquito or Man?,* 188.

43. Boyce, *Yellow Fever Prophylaxis in New Orleans*, 6–7; Joseph Goldberger Papers, Folder 1, Southern Historical Collection, University of North Carolina at Chapel Hill.

44. Walter Wyman to Joseph Goldberger, March 22, 1904, Goldberger Papers, Folder 3.

45. Gorgas, *Sanitation in Panama*, 138; Philippe Bunau-Varilla, *Panama: The Creation, Destruction, and Resurrection* (New York: McBride, Nast, & Co., 1914), 542.

46. Patrick Manson, *The Relation of the Panama Canal to the Introduction of Yellow Fever into Asia* (London: Bedford Press, 1903), 5; Douglas Haynes, *Imperial Medicine: Patrick Manson and the Conquest of Tropical Disease* (University of Pennsylvania Press, 2001), 2.

47. Manson, *Relation of the Panama Canal to the Introduction of Yellow Fever*, 10–11.

48. Ibid., 10.

49. Gorgas, *Sanitation in Panama*, 139.

50. James W. Porch and Fred Muller, "Panama via New Orleans," *Report of the Board of Trade Committee*, November–December, 1904, 3–4.

51. Ibid., 3–4.

52. Ibid., 8, 11.

53. Boyce, *Mosquito or Man?*, 185; Gorgas, *Sanitation in Panama*, 185.

54. Boyce, *Yellow Fever Prophylaxis in New Orleans*, 6–7; Goldberger Papers, Folder 1.

55. A. H. Glenman (Acting Surgeon General) to Joseph Goldberger, July 26, 1905, Goldberger Papers, Folder 4.

56. Boyce, *Yellow Fever Prophylaxis in New Orleans*, 18–19.

57. Ibid., 50–60.

58. "Swamp" mosquitoes were usually anophelenes that carried malaria, which remained endemic in Louisiana until the 1950s. See Boyce, *Yellow Fever Prophylaxis in New Orleans*, 38.

59. Boyce, *Yellow Fever Prophylaxis in New Orleans*, 38, 46–49.

60. Ibid., 40, 53, 58.

61. Boyce, *Mosquito or Man?*, 178–80.

62. Rubert Boyce, *Health Progress and Administration in the West Indies* (New York: E. P. Dutton and Co., 1910), 162; *Mosquito or Man?*, 192–93.

63. Boyce, *Health Progress and Administration in the West Indies*, 49.

64. Ibid.

65. Boyce, *Mosquito or Man?*, 192–95.

66. Haynes, *Imperial Medicine*, 176; David Arnold, ed., *Warm Climates and Western Medicine: The Emergence of Tropical Medicine, 1500–1900* (Amsterdam: Rodopi, 1996), 4–6; Stepan, *Picturing Tropical Nature*, 158–70; Conevery Bolton Valenčius, *The Health of the Country: How Americans Understood Themselves and Their Land* (New York: Basic Books, 2002), 231–32; Cecilia Elizabeth O'Leary, *To Die For: The Paradox of American Patriotism* (Princeton, N.J.: Princeton University Press, 1999), 116.

67. Arnold, ed., *Warm Climates and Western Medicine*, 5.

68. Michael Worboys, "Germs, Malaria and the Invention of Mansonian Tropical Medicine: From 'Diseases in the Tropics' to 'Tropical Diseases,'" in *Warm Climates and Western Medicine,* ed. Arnold, 193–94.

69. Haynes, *Imperial Medicine,* 176.

70. Ibid., 3.

71. Julyan G. Peard, "Tropical Medicine in Nineteenth-Century Brazil: The Case of the 'Escola Tropicalista Bahiana,' 1860–1890," in Arnold, ed., *Warm Climates and Western Medicine,* 108.

72. Ibid., 108–11; see also Julyan G. Peard, *Race, Place, and Medicine: The Idea of the Tropics in Nineteenth-Century Brazilian Medicine* (Durham, N.C.: Duke University Press, 1999).

73. For an overview of the reemergence of yellow fever in Africa after 1950, see the epilogue in Willoughby, *Yellow Fever, Race, and Ecology in Nineteenth-Century New Orleans.*

74. Studies of unintended consequences of malaria-control campaigns might be useful in determining how to evaluate the effects of campaigns to eradicate yellow fever. James L. A. Webb has shown how the irregular use of antimalarials and insecticides resulted in the development of resistance among malaria parasites and *Anopheles* mosquitoes. Gaps in malaria-control efforts caused the loss of acquired immunities among local populations, resulting in higher mortality and morbidity rates among African adults. See James L. A. Webb, *The Long Struggle against Malaria in Tropical Africa* (Cambridge, U.K.: Cambridge University Press, 2014), 114–32.

THE COMPLETE STORY
OF THE GALVESTON HORROR
Trauma, History, and the Great Storm of 1900

ANDY HOROWITZ

The darkest horror of American history has fallen on our southern coast.
—William J. McGee, "The Lessons of Galveston," 1900

On September 8, 1900, a hurricane in the Gulf of Mexico pushed a fifteen-foot-tall wall of water onto the island of Galveston, Texas. The highest point in the city of roughly forty thousand residents was nine feet above sea level. At least six thousand and perhaps as many as ten thousand people died in the flood, making the unnamed 1900 storm the deadliest hurricane in the history of the United States.[1]

New York Herald reporter John Coulter quickly published an account of what, supposedly, had transpired. His book, *The Complete Story of the Galveston Horror,* promised readers "Incidents of the awful Tornado, Flood and Cyclone Disaster; Personal Experiences of Survivors; Horrible Looting of Dead Bodies and the Robbing of Empty Homes; Pestilence from so many Decaying Bodies Unburied; Barge Captains Compelled by Armed Men to Tow Dead Bodies to Sea; . . . Tales of the Survivors from Galveston; Adrift all Night on Rafts; Acts of Valor; United States Soldiers Drowned; Great Heroism; Great Vandalism; Great Horror; A Second Johnstown Flood, but worse; Hundreds of Men, Women and Children Drowned." "There was no way of escape," Coulter concluded his dystopian litany, "only Death! Death! Everywhere!"[2] This essay is an attempt to make sense of a small percentage of those deaths: not the ones caused by the wall of water, but rather those caused by the white vigilantes and militiamen who boasted, after the storm, of executing dozens of African Americans for looting in Galveston's devastated streets.

Descriptions of people looting the ruined city appear throughout oral history interviews, survivors' memoirs, and the sensationalist books like Coulter's published soon after the flood. "Following the catastrophe," Coulter reported, for instance, "Galveston was practically in the hands of thieves, thugs, ghouls, vampires, and bandits, some of them women, who robbed the dead, mutilated the corpses which were lying everywhere, ransacked business houses and residences and created a reign of terror." Their "reign" continued for two days, Coulter wrote, until Galveston mayor Walter C. Jones declared martial law and called up the local militia. "With commendable promptness the regulars put the ghouls under arrest," Coulter asserted, "and without ceremony shot every one of them."[3] Although Coulter's language is particularly dramatic, and estimates vary as to the extent of the violence, almost all of the accounts agree that the looters allegedly causing disorder were black and the shooters restoring order were white.

Most historians writing about the Galveston flood have dismissed these reports. "Nothing of the sort happened," Erik Larson asserts with definitive succinctness in *Isaac's Storm,* his best-selling history of the disaster.[4] The widely varying estimates of how many looters were shot, the lack of hard evidence that looting occurred in the first place, the folkloric language of "vampires" and "ghouls," and the racism that pervades the accounts all suggest that the looting and executions might have existed only—or at least primarily—in rumor and fantasy. That may be why most historians have focused on the Galveston flood as "the worst natural disaster in the history of the North American continent," and as a catalyst for the conception of the "commission" form of government, a signal innovation in the rise of the Progressive New South.[5] Fit together by contemporary observers and later historians alike, the two tropes structure a narrative of "creative destruction," in which a chaotic old order is cleared away and a new, more efficient one emerges.[6]

Taken as fact or folklore, however, the fantasy of racial violence that followed the storm challenges the central themes that pervade many histories of Galveston specifically and disaster narratives generally. Though a growing number of scholars argue that disasters are best understood as historical processes, many writers continue to narrate so-called natural disasters as acute events that erupt in a catastrophic instant. In these tellings, outsize pathos seems to break time and fracture chronology. A recent journal article described Galveston and several other disasters, for example, as "Moments When Time Stood Still."[7] The title of an excellent collection of primary sources from Galveston rhetorically confines the event to lasting "Through A Night of Horrors."[8] In the field of disaster studies—the definition of the field itself suggesting that disasters are a discrete, coherent category standing apart from the normal flow of life—this ahistorical view of disasters as extraordinary events helps to undergird the prominent theory that disasters give rise to utopian communities. "It has long been noted," the sociologist Charles Fritz wrote in an influential article in 1960, "that disasters unify societies."[9] The story is as old as Noah's flood: temporarily washing away divisive cultural and historical norms, Fritz and his followers have argued, disasters allow humanity's basic altruism to emerge. "The prevalent human nature in disaster," the author Rebecca Solnit wrote in *A Paradise Built in Hell: The Extraordinary Communities That Arise in Disaster,* "is resilient, resourceful, generous, empathic, and brave."[10]

The lurid accounts of the mayhem and murder that purportedly overwhelmed Galveston in the two days following the storm, however, make clear the ways in which the disaster was not fully meteorological, and did not pass in just one night of horrors. These accounts lurk beneath and within the twin historiographical frames of an unprecedented natural disaster and a heroically rational response, and they undermine hopes that a baptism by tidal wave might wash away the sins of the past. The stories of these deaths suggest that an event famously extraordinary ("the most appalling calamity of modern times," in the words of one ambitious author) is thoroughly imbricated in its ordinary historical context, and that at least some of the traumatizing agency usually assigned to the meteorological coincidence of wind and water belongs rather to the Texas vigilantes and the writers who broadcast their spectacular claims across the country.[11] The accounts of looting and lethal responses to it, in other words, illuminate how the catastrophe of the Galveston flood was neither natural nor fleeting. Rather, it was bound up in the ongoing tragedy of racial terror that accompanied the rise of Jim Crow at the turn of the twentieth century. Like history itself, disasters exist in time. Their trauma blurs the boundary between the acute and the chronic.

"The Negroes Were Shot Down"

The storm surge precipitated by the unnamed hurricane of September 8, 1900, killed at least one out of every six residents of Galveston. The extent of the devastation struck writers as entirely without precedent. "Such a night of horror as the unfortunate inhabitants were compelled to pass," Coulter wrote, "has fallen to the lot of few since the records of history were first opened." The damage, he asserted, "had no parallel in history."[12] When the clouds cleared from the sky on September 9, the people who had managed, somehow, to survive the flood found their city in ruins, "a scene of suffering and devastation," *National Geographic* reported later, "hardly paralleled in the history of the world."[13] It was "a city of . . . stunned and stricken people . . . all searching for their friends among the slain," the *Galveston Tribune* reporter Clarence Ousley wrote, "tearless but bleeding at the heart . . . a city of wrecked homes and streets choked with debris sandwiched with six thousand corpses."[14] Most of Galveston's buildings had collapsed. Thousands of friends, neighbors, and family members lay dead in the streets.

In oral histories and survivor memoirs, white Galvestonians recounted stories of African Americans pilfering jewelry from these corpses. Harry Maxon wrote in a memoir that after the storm he saw a "Negro man" with "a flour sack filled with mostly jewelry taken from dead bodies that were floating around everywhere."[15] "There was one man that . . . had his pockets full of ears and fingers," Ellen Edwards Nilson recalled in an interview. "He didn't take the time to take the earrings out of ears, he just cut a piece off. Of course by that time the fingers would be swollen. He just cut the fingers off and stuffed them in his pocket."[16]

Published accounts offered even more shocking versions of the same stories, joining Galveston with the 1898 Johnstown Flood in what amounted to a burgeoning genre of disaster melodrama. In *Galveston:*

The Horrors of a Stricken City, journalist Murat Halstead reported, "The ghouls were holding an orgie [*sic*] over the dead. The majority of these men were negroes." And "not only did they rob the dead, but mutilated bodies in order to secure their ghoulish booty." Halstead indulged in pornographic descriptions of violence, presumably hoping his scandalous images would attract readers. "A party of ten negroes was returning from a looting expedition," he wrote. "They had stripped corpses of all valuables, and the pockets of some of the looters were fairly bulging out with fingers of the dead, which had been cut off because they were so swollen the rings could not be removed."[17] He continued, "During the robbing of the dead not only were fingers cut off, but ears were stripped from the head in order to secure jewels of value."[18]

"Human nature at its worst has had opportunity for the display of its meanest passions," Paul Lester wrote in his similarly garish *Great Galveston Disaster.* "Looters and vandals have ignored all moral restraints." Lester described one looter as having "in his pocket twenty-three human fingers with costly rings on them."[19] Halstead cast the looting as a dark "orgie," but Lester called it spectacular revelry. "Negro looters," he wrote, "held high carnival."[20]

Most scholars agree that people tend not to loot during disasters. They may engage in what amounts to foraging, or requisitioning basic necessities, but "the basic proposition advanced from the earliest studies," the disaster sociologist Enrico Quarantelli has written, is that looting for the sake of capital acquisition is "very rare and in many cases almost nonexistent."[21] Fritz and Solnit emphasize this point in their work.

Disasters, Fritz claimed, stand outside of the conflicts of history. They "provide a temporary liberation from the worries, inhibitions, and anxieties associated with the past and the future," he asserted, "because they force people to concentrate their full attention on immediate moment-to-moment and day-to-day needs." By removing people from history, Fritz argued, disasters also remove people from culture. "The blanking out of the past and future frames of reference," he wrote, leads to "the emergence of new social norms and values that are fitted to present realities." Disasters "provide an arena for the observation of peculiarly human (as opposed to culturally unique) behavior," Fritz explained, because "culturally derived discriminations and social distinctions tend to be eliminated in disaster because all groups and statuses in the society are indiscriminately affected." Echoing Jean-Jacques Rousseau's notion that humans, in a state of nature, are fundamentally cooperative and decent—and writing in a way that may have comforted readers anxious about the possibility of atomic war in the early 1960s—Fritz offered an optimistic view of human resilience. He filed the new social norms that arise during catastrophe under the happy category of "therapeutic adjustments to disaster."[22]

An analytical reading of the social science literature, an optimistic reading of human nature, or a critical reading of the racist mythology that structures these narratives of African American lawlessness and inhumanity: all three approaches suggest that people of color lucky enough to survive the Galveston flood—deprived of shelter and sustenance, suffering the loss of their own family members and neighbors—most likely did not engage promptly in an orgy or carnival of severing fingers and ears from bloated corpses in pursuit of pilfered jewelry. Nonetheless, some influential white Galvestonians believed that they would.

Flood-ravaged Galveston saw no liberation from history. Across the city, the disaster aggravated rather than suspended preexisting social conflicts. Lloyd Fayling, for example, the "district manager for a New England publishing syndicate" in Galveston, spent the storm indulging his own racialized fears. Fayling had been a deputy U.S. marshal in Chicago during the 1894 railroad strike, presumably helping to break the strike; he had been a newspaper reporter in Cuba before the Spanish-American War and might have fought in the war itself.[23] He rode out the flood in a house with forty-three other "refugees," all of whom were grateful for his help, he reported, "except about fifteen Negroes and hoboes who spent the night under arrest." In a memoir, Fayling wrote that he overheard these men plotting to commandeer a boat to escape the flood, but "I fortunately had a six-shooter and a Winchester handy, and put them under arrest at pistol point." Fayling and an assistant "made those Negroes' lives most unhappy ones." After the storm surge subsided, Fayling led his prisoners to City Hall but found the Galveston lockup destroyed. He released them with a "lecture and a stern warning." But the trouble, as Fayling perceived it, was only beginning. As he walked through the city, he saw society coming unbound. "Everything was chaos," he observed. "All sorts of suspicious looking people were crawling over heaps of rubbish and going in and out of stores in the most suspicious manner," he wrote. With a foreboding air of inevitability he concluded, "The looting had already begun."[24]

Fearing lawlessness and disorder, Fayling fashioned himself an instrument of authority. He managed to track down the Galveston chief of police, Ed Ketchum, and received a commission as "Sargent [*sic*] of Police." That status, whatever its actual legal significance, empowered Fayling to embark on a crusade to restore order. He quickly started recruiting—and impressing—men from the streets for service in an ad hoc militia.[25] The reporter Ousley called Fayling's a "quasi-military organization" that "was without legal authority, state or federal," but nonetheless "managed by display of uniform and bayonet to overawe such desperation as might breed."[26] Evidently stoking old fears but now newly licensed to act on them, Fayling believed "the only salvation of the town lay in getting it under martial law as quickly as possible." Fayling ordered his men, first, to realize a dream of Galveston's Progressive Era reformers by closing all the saloons, and second, to "shoot anyone caught looting the dead or desecrating corpses in any way. If anyone resists your authority," Fayling ordered, "shoot."[27]

To Fayling, the entire city appeared chaotic: even the animals seemed to crave the restoration of order. "Horses and cattle were straying in every direction in the streets," he observed, employing the language of slavery, "ownerless, and so anxious for human company that they would run up to every passer-by as if looking for their masters." In Fayling's account, nature itself assented to his authority. "The horses were glad indeed to come at the order," he affirmed.[28]

In his memoir, Fayling does not record shooting any looters, emphasizing instead his own powerful command of the city. "We had the city under complete control from the first," he asserted. "Order was absolutely maintained." He did boast of many arrests. "We drove hundreds of Negroes at the bayonet point to assist in the work of burning and loading the dead on barges for sea burial," he reported too. Keeping them plied with whiskey from the saloons he had recently closed, Fayling forced these men to engage in the terrible work of disposing of the dead.[29]

Whether or not Fayling ever fired a shot, he was keenly aware of how to use the threat of violence to control others. The "hundreds of Negroes" that he impressed into service were loath to carry corpses to the funeral pyre. "'For heavens sake don't make me do that!'" Fayling reports them crying. "'I won't go, you can shoot me if you want to, but I will not and I can not.'" Fayling asserted, "Our only answer was 'Load with ball cartridge, take aim—' and fortunately we never had to go any further. They always threw up their hands and went to work." Later, Fayling reflected, "I do not know whether I would have shot them or not. But . . . I think I would."[30]

Fayling also used the threat of execution to keep his white militiamen at attention. He stationed sentries on street corners throughout Galveston and warned them against falling asleep at their posts. He proudly described "a superstition among them which I carefully fostered, that I was going to shoot anybody who went to sleep." He keenly understood the power of such terror. "Of course I would not really have done this," he wrote, clarifying that he would have shot an African American man but not a white man, "but it was just as well to let them think that I would."[31]

If the ad hoc militia under Fayling's command did not use lethal force, many published reports suggest that other vigilantes did.[32] The accounts of how many men were killed vary widely. "Ten ghouls—eight negroes and two whites—were caught after robbing bodies," Halstead wrote of one instance. "Their pockets were filled with fingers and ears, cut from corpses. These pieces of flesh bore rings." In Halstead's account, justice for the looters was distributed along racial lines: "the negroes were shot down," he wrote, while revealing nothing about the fate of the so-called white ghouls.[33] "Martial law reigns here," Halstead noted approvingly. "Fiends who, like buzzards, thrive at such times as this, are shown no mercy; are given no trials. The orders are to shoot them down and these orders are obeyed." Halstead estimated, "In all about fifty ghouls, despoilers of the dead, have been shot down, and a negro who attacked a woman has been killed."[34] Coulter's estimate was higher. "Tuesday night ninety negro looters were shot in their tracks by citizens guards," he wrote. "One of them was searched and $700 found, together with four diamond rings and two water-soaked gold watches. The finger of a white woman with a gold band was clutched in his hands."[35] Even though the narrative suggests that the suspected looter was searched only after he was shot, the careful evocation of deeply held white southern fears of African American men despoiling white women is meant to justify the action.[36]

Some white Galvestonians doubted the truth of the more spectacular rumors, and their oral histories and memoirs suggest that fewer people actually were executed for looting. On September 10, John Blagden, a forecaster temporarily reassigned to Galveston from his regular U.S. Weather Bureau post in Memphis, wrote to his family in Minnesota, "The city is under military rule and the streets are patrolled by armed guards. They are expected to . . . shoot at once any one found pilfering." Blagden wrote that he had heard that four men had been shot that day, but he was unconvinced. "I do not know how true it is for all kinds of rumors are [afloat] and many of them are false."[37] James Brown, an English emigrant who had just arrived in Galveston from Flatonia, Texas, wrote to his sisters and cousins a month after the flood, "About 20 men was shot dead for robbing the dead of rings and jewelry."[38]

Ousley, the local reporter, offered what he meant as a sober corrective. "There was some looting," he allowed. "In every community there are ghoulish natures." In Galveston, "the remedy was swift and effective, as the situation required." Discounting what he called exaggerations, Ousley wrote that while "reports at the time represented that as many as seventy-five ghouls were shot in their tracks," in fact, "diligent inquiry fails to discover conclusive proof of one-tenth the number. It may be safely put down that if any were killed the number will not exceed a half-dozen." Nonetheless, he acknowledged the power of the rumors. "At any rate," Ousley concluded, "the reports and display of force served a good purpose," implying that they prevented people from looting who might otherwise have done so.[39]

Just as there are many reasons to doubt the veracity of the looting rumors, there is similar cause to question the dramatic stories of vigilante response. In this case, though, the scholarly literature on disaster is less conclusive. Sociologists have developed the concept of "elite panic," asserting that upper-class people tend to perceive disorder during disasters more than the poor do.[40] In a way that resonates with Fayling's account of his own actions, Solnit explains, "The elite often believe that if they themselves are not in control, the situation is out of control, and in their fear take repressive measures that become secondary disasters."[41] The notion that class distinctions may account for different responses to disaster may be reasonable, although it sits uncomfortably with the theory that disasters serve to erase social distinctions.

The facts of history suggest that, while widespread looting in Texas would have represented a break from the past, lynching—which is how many observers interpreted the vigilante response—would not. "Troops patrolling the island did not hesitate to kill every one of the vandals caught in the commission of his infamous work," Coulter wrote. "Public opinion sustained this prompt style of punishment. It was a species of Southern lynching to which no objection was ever raised."[42] Elsewhere, Coulter related the story of a Mexican man who appeared to be looting. "With the sight of these evidences of crime before them," Coulter wrote, observers "seemed to go mad, and with cries of 'Lynch him!' 'Burn him!' made for the unfortunate wretch."[43] Texans had heard such cries before: over a hundred African Americans were lynched in the United States in 1900, and more than twenty people were lynched in Texas alone every year during the surrounding decade.[44]

The association of lethal martial law with lynching connects the exigencies of the Galveston catastrophe to the broader context of turn-of-the-century Texas. In *The Great Galveston Disaster,* Lester described one Galvestonian, for instance, insisting, "The negroes should be sent to the cotton fields of north Texas. Those who will work can be kept there, but the others should be sent away just as soon as possible, for they merely eat up the supplies and are a constant menace." Ideas such as these circulated before the storm, but the speaker now couched them as a form of emergency response. "They should either be killed or made to get out," he asserted, "for one or the other is the grim necessity of the situation."[45]

The New Order

Mayor Jones believed that order, so-called, was restored rather quickly in the context of the immediate crisis. He relieved Fayling on September 11 and put city residents under the command of General Thomas Scurry, the adjutant-general of the Texas State Volunteer Guard. The city of Galveston remained under martial law until September 20.[46] Eager white reformers and businessmen then sought to bring order to Galveston's governance writ large.

In most accounts, the heroic recovery—this disaster's silver lining— was the imposition of the "commission form" of government. Reformers designed the commission system to dismantle what they viewed as a corrupt and inefficient political machine. Building on efforts begun in the 1890s, reformers drew up a proposal for a new city charter in November 1900, little more than a month after the storm. The Texas legislature approved the charter, with slight modifications, in July 1901. The new charter called for a government of five city commissioners: three appointed by the governor and two elected at-large by Galveston voters. Each of the five commissioners was responsible for a specific city department: "finance and revenue, police and fire, waterworks and sewerage, streets and public improvements. The entire commission, sitting as a body, would make policy decisions." The new commissioners took office in September 1901.[47] The land speculator and one-time postmaster of Galveston, Edmund Cheesborough, described the Galveston City Commission as a "Board of Municipal Directors," "composed of five practical business men, each fully recognizing the fact that economy and business methods, not politics, should be employed in transacting the business affairs of the city."[48]

White reformers believed that the city commission system transferred control of municipal functions to rational, efficient managers, creating a powerful bureaucracy competent enough to rebuild the city and prepare for an ever-brighter future. "For a time," the Progressive

Era historian Charles Beard wrote of Galveston after the storm, "the local government was paralyzed, because the problems connected with the reparation of the ruin were too much for the old political machine which had control."[49] The commission, Beard believed, insulated government from irrational factional interests. Tying together the tropes of an unpredictable natural disaster with a progressive narrative and a belief in altruistic human nature, another historian of Galveston described the advent of the commission system this way: "What happened in Galveston happened because young progressives . . . were basically smart, honest, and altruistic, and because they were in the right place at the right time."[50] "No single movement of reform in our governmental methods has been more significant," Woodrow Wilson asserted in 1911, when he was governor of New Jersey, "than the . . . commission form of government."[51] The iconic achievement of Galveston's post-storm government was the construction of a seawall meant to protect the island from future storms: a concrete bulwark against disorder and unruly nature.

With the majority of its members appointed by the governor, the commission was less democratic than the system it replaced. African Americans suffered the biggest loss. In the new system, the two members not appointed by the governor were elected "at-large," ensuring that African American voters would always be outnumbered in local elections. "There is a most ridiculous way of the operation of things for the betterment of a future Galveston," the African American Galveston newspaper the *City Times* argued in an editorial on September 26, 1900, as the plans for the new city charter took shape. "The colored man is good enough to save the lives of the little white babes, white women and even men," the editorial continued, suggesting an alternate version of what transpired during the hurricane. "Good enough to . . . render noble assistance in every particular to help uplift the stricken and the sinful Galveston to a new future." Nonetheless, the editorial protested, "in all of that he has not been good enough to even be represented as a committeeman."[52]

The Galveston plan for a city commission—supposedly inspired by an act of God, the Galveston disaster—thus stands alongside the imposition of the poll tax in 1901 and the white primary in 1923 as a key moment in Texas's disfranchisement of African Americans.[53]

Conclusion

Trauma is a product of history: it works not in the catastrophic instant but across the stretches of time that give life its meaning. The same is true of a disaster: its causes and consequences take time to unfold. "Disasters," the sociologist Fritz wrote in 1961, "are not like the crises of everyday life."[54] But the story of the Galveston storm shows how blurred the boundaries are between the acute and the chronic, between discrete events and ongoing processes. The insistence on viewing a disaster as comprising only "a night of horrors," a calamity out of time, supposes that there exists some other reality to return to afterwards. For many people, in many places, the hallmarks of what we call disasters are, for them, characteristic of everyday life.

With that in mind, in *Everything in Its Path,* his extraordinary study of the 1972 Buffalo Creek flood, sociologist Kai Erikson proposed a new way of defining the relationship between disaster and trauma. "Instead of classifying a condition as a trauma because it was induced by a disaster," he wrote, suppose "we would classify an event as disaster if it had the property of bringing about traumatic conditions." Such a redefinition, Erikson observed, would "suggest that there are any number of happenings in human life that seem to produce the same effects as a conventional disaster without exhibiting the same physical properties."[55] The point of thinking about Galveston in the terms that Erikson proposed would not be to diminish the shocking calamity of the fifteen-foot-tall wall of water that came from the Gulf of Mexico and killed over six thousand people in a single blow, but rather to expand our sense of the disaster—to complete the story of the Galveston horror—by adding to it the disfranchisement that announced the New South, the extralegal spectacle of lynching masked as the law itself, the violent fantasies at the heart of white supremacy, and the ongoing human disaster of racialized terror.

Foregrounding trauma as a category of experience can help to reframe the fractured history and memory of the Galveston disaster.[56] In the existing historiography of Galveston, the storm is the disaster, one that brings a cathartic breakthrough. But seeing racism, inequality, and violence as the disaster—thinking of disaster as a chronic human process rather than an acute wound from nature—reveals the storm to have been an occasion not for what the scholar of trauma Dominic LaCapra calls "working through," but rather "acting out": instead of allowing white Galvestonians to transcend their history of violence against African Americans, the storm seemed to authorize them to further enact and reenact the imposition of suffering.[57] Viewed through the lens of trauma studies, the Galveston storm is a "screen memory," in which meteorology obscures sociology, and natural forces stand in for the work of human hands. Attributing the explosion of violence that followed the storm to the storm itself exculpates the racist vigilantes by naturalizing their behavior, and casts their behavior as inevitable, rather than a conscious choice.[58]

The white militiamen who fantasized about killing African Americans in the streets to restore order, and the reporters who broadcast their stories around the country, must have believed in the righteousness of those imagined acts of heroism. "Political rhetoric can license people to do evil in the name of good," observed historian Glenda Gilmore in reference to an 1898 racial massacre in North Carolina, another instance of the wave of racial violence that swept across the South at the turn of the century. Those vigilantes and their biographers must also have known something about the way the stories from Galveston—real or imagined—would resound and become their own kind of truth and do their own kind of violence. The disastrous and traumatizing reality that such actions might be possible and were desired by many would endure long after the floodwaters receded and the dead were cleared from the streets. "Murder's best work is done after the fact," Gilmore wrote, "when terror lives on in memory."[59]

In 2008, more than a century after the unnamed storm described here, another hurricane made landfall on the east coast of Texas. Across the bay from Galveston, two white men readied themselves for the disaster they feared would loom after the storm. They loaded their shotguns and spray-painted a warning on a piece of plywood. The words they wrote functioned like stage directions for a play the two men were sure they had seen before. "The repetition at the heart of catastrophe," the scholar Cathy Caruth asserts, "emerges as the unwitting reenactment of an event that one cannot simply leave behind."[60] The men's sign proclaimed: "U LOOT WE SHOOT."[61]

NOTES

An Archibald Hannah Jr. Fellowship in American History from the Beinecke Rare Book and Manuscript Library supported the research for this project, and I am grateful to George Miles for helping me to identify many of the primary sources. Yoav Di-Capua offered a particularly useful critique of an earlier draft. I remain indebted to him, Seth Garfield, and the other scholars who participated in the Trauma and History Conference at the University of Texas–Austin's Institute for Historical Studies. Sarah Gray, too, offered crucial insights.

1. The most prominent recent account of the storm is Erik Larson, *Isaac's Storm: A Man, a Time, and the Deadliest Hurricane in History* (1999; New York: Vintage, 2000).

2. John Coulter, ed., *The Complete Story of the Galveston Horror* (n.p.: E. E. Sprague, 1900), n.p.

3. Ibid., 43 (first quotation) and 45 (second quotation).

4. Larson, *Isaac's Storm*, 242.

5. Elizabeth Hayes Turner, *Women, Culture, and Community: Religion and Reform in Galveston, 1880–1920* (New York: Oxford University Press, 1997), 187 (for "natural disaster"); Patricia Bellis Bixel and Elizabeth Hayes Turner, *Galveston and the 1900 Storm: Catastrophe and Catalyst* (Austin: University of Texas Press, 2000).

6. "Long before the end of the decade all the scars of the great storm will have been erased," A. H. Belo wrote, for example, "and a new and strong Galveston will replace that so ruthlessly destroyed by the recent storm." He continued, "the clouds of the gloomy present will soon disappear to reveal a future full of hope and promise" ("Galveston—What It Was, and What It Will Be," *Harper's Weekly,* October 6, 1900, 935). For "creative destruction," see Joseph Schumpeter, *Capitalism, Socialism & Democracy* (1942; London: George Allen & Unwin, 1976), 83. See also Kevin Rozario, *The Culture of Calamity: Disaster and the Making of Modern America* (Chicago: University of Chicago Press, 2007), esp. 84–86.

7. Patricia D'Antonio and Jean C. Whelan, "Moments: When Time Stood Still," *American Journal of Nursing* 104, no. 11 (November 2004): 66–72.

8. Casey Edward Greene and Shelly Henley Kelly, eds., *Through a Night of Horrors:*

Voices from the 1900 Galveston Storm (College Station: Texas A&M University Press, 2000).

9. Charles Fritz, "Disaster," in *Contemporary Social Problems: An Introduction to the Sociology of Deviant Behavior and Social Disorganization*, ed. Robert K. Merton and Robert A. Nisbet (New York: Harcourt, 1960), 651–94 (quotation on 683).

10. Rebecca Solnit, *A Paradise Built in Hell: The Extraordinary Communities That Arise in Disaster* (New York: Viking, 2009), 8.

11. Paul Lester, *The Great Galveston Disaster: Containing a Full and Thrilling Account of the Most Appalling Calamity of Modern Times* (n.p.: Horace C. Fry, 1900).

12. Coulter, ed., *The Complete Story of the Galveston Horror*, 33.

13. W. J. McGee, "The Lessons of Galveston," *National Geographic* 11, no. 10 (1900): 377.

14. Clarence Ousley, *Galveston in Nineteen Hundred* (Atlanta: William C. Chase, 1900), 23.

15. Qtd. in Greene and Kelly, eds., *Through a Night of Horrors*, 134.

16. Ellen Edwards Nilson, interviewed by Glen Echols, no date [the interview was conducted before her death in 1966], 13, Rosenberg Library, Galveston and Texas History Center, Galveston (hereafter GTHC), www.gthcenter.org/exhibits/storms/1900/Oral hist/Nilson.htm. See also Greene and Kelly, eds., *Through a Night of Horrors*, 178.

17. Murat Halstead, *Galveston: The Horrors of a Stricken City* (n.p.: H. L. Barber, 1900), 99.

18. Ibid., 100.

19. Lester, *The Great Galveston Disaster*, 61.

20. Ibid., 65.

21. Quarantelli specifies that he is referring to "American-type communities," though he does not clarify precisely what this means. See Enrico L. Quarantelli, "Conventional Beliefs and Counterintuitive Realities," *Social Research* 75, no. 3 (Fall 2008): 873–904 (quotation on 883). For a concise overview of the divergence between the myth and the reality of disaster response, see Kathleen Tierney, Christine Bevc, and Erica Kuligowski, "Metaphors Matter: Disaster Myths, Media Frames, and Their Consequences in Hurricane Katrina," *Annals of the American Academy of Political and Social Science* 604, no. 1 (March 2006): 57–81, esp. 57–61.

22. See, for example, Jean-Jacques Rousseau, "Discourse on Inequality" (1754). Fritz, "Disaster," 688 (for "liberation" and "blanking out"), 685 (for "culturally derived"), 655 (for "an arena"), and 688 (for "therapeutic").

23. Greene and Kelly, eds., *Through a Night of Horrors*, 75.

24. Fayling, untitled account of his experiences in the Galveston storm (1905), 5–6, Lloyd R. Fayling Papers, MSS #80–0021 (GTHC), www.gthcenter.org/exhibits/storms /1900/Manuscripts/Fayling/index.html. See also Greene and Kelly, eds., *Through a Night of Horrors*, 80–81.

25. Fayling, untitled account, 6. See also Greene and Kelly, eds., *Through a Night of Horrors*, 83.

26. Ousley, *Galveston in Nineteen Hundred*, 250.

27. Fayling, untitled account, 9. See also Greene and Kelly, eds., *Through a Night of Horrors*, 83.

28. Fayling, untitled account, 10. See also Greene and Kelly, eds., *Through a Night of Horrors*, 84.

29. Fayling, untitled account, 11. See also Greene and Kelly, eds., *Through a Night of Horrors*, 85.

30. Fayling, untitled account, 11–12. See also Greene and Kelly, eds., *Through a Night of Horrors*, 86.

31. Falying, untitled account, 10. See also Greene and Kelly, eds., *Through a Night of Horrors*, 85.

32. A lethal response to looting likely would have been illegal because of the "rule, pervasive at least in Anglo-American jurisprudence, that deadly force may not be used in defense of property, even when it is the only means available to prevent the loss" (Stuart P. Green, "Looting, Law, and Lawlessness," *Tulane Law Review* 81, no. 4 [March 2007]: 1129–74 [quotation on 1170]).

33. Halstead, *Galveston*, 96.

34. Ibid., 96–99.

35. Coulter, ed., *The Complete Story of the Galveston Horror*, 57.

36. See, for instance, Glenda Elizabeth Gilmore, "Murder, Memory, and the Flight of the Incubus," in David S. Cecelski and Timothy B. Tyson, eds., *Democracy Betrayed: The Wilmington Race Riot of 1898 and Its Legacy* (Chapel Hill: University of North Carolina Press, 1998): 73–94.

37. John Blagden, "Letter to family in Duluth, Minnesota," September 10, 1900, p. 5, John D. Blagden Papers, MSS #46-0006 (GTHC), www.gthcenter.org/exhibits/storms /1900/Manuscripts/Blagden/index.html.

38. James Brown to his sisters and cousins, October 7, 1900, p. 6, James Brown– Winnifred B. Clamp Letters, MSS #85-0013 (GTHC), www.gthcenter.org/exhibits /storms/1900/Manuscripts/Brown_Clamp/index.html. See also Greene and Kelly, eds., *Through a Night of Horrors*, 65.

39. Ousley, *Galveston in Nineteen Hundred*, 37–38.

40. Qtd. in Solnit, *A Paradise Built in Hell*, 127.

41. Solnit, *A Paradise Built in Hell*, 21.

42. Coulter, ed., *The Complete Story of the Galveston Horror*, 271.

43. Ibid., 113.

44. Monroe N. Work, *Negro Year Book* (Tuskegee, Ala.: Tuskegee Institute, 1937), 156; John R. Ross, "Lynching," *The Handbook of Texas Online* (Austin: Texas State Historical Association, 2010), www.tshaonline.org/handbook/online/articles/jg101 (accessed May 5, 2015).

45. Lester, *The Great Galveston Disaster*, 204.

46. Ousley, *Galveston in Nineteen Hundred*, 251.

47. Bixel and Turner, *Galveston and the 1900 Storm*, 92–93 (quotation on 92).

48. Governed by business interests and ideas, the commission further fits the overarching narrative of "creative destruction" and perhaps even Naomi Klein's concept of "disaster capitalism." See George Kibbe Turner, "Galveston: A Business Corporation," *McClure's*, October 1906, qtd. in Bradley R. Rice, "The Galveston Plan of City Government by Commission: The Birth of a Progressive Idea," *Southwestern Historical Quarterly* 78, no. 4 (April 1975): 408. See also: Naomi Klein, *The Shock Doctrine: The Rise of Disaster Capitalism* (New York: Picador, 2008); E. R. Cheesborough, "Galveston's Commission Plan of City Government," *Annals of the American Academy of Political and Social Science* 38, no. 3 (November 1911): 221–30 (quotation on 221).

49. Charles A. Beard, *American City Government: A Survey of Newer Tendencies* (New York: Century Co., 1912), 93, qtd. in Rice, "The Galveston Plan of City Government by Commission," 378.

50. Gary Cartwright, *Galveston: A History of the Island* (Fort Worth: TCU Press, 1991), 185–86.

51. Woodrow Wilson, *Galveston Daily News,* May 22, 1911, qtd. in Rice, "The Galveston Plan of City Government by Commission," 365.

52. "Suggestions and Advice about Conditions of the Poor, Suffering People," *Galveston City Times,* September 29, 1900, 1.

53. Bixel and Turner, *Galveston and the 1900 Storm,* 149.

54. Fritz, "Disaster," 683.

55. Kai T. Erikson, *Everything in Its Path: Destruction of Community in the Buffalo Creek Flood* (New York: Simon & Schuster, 1976), 254.

56. I rely heavily here on Yoav Di-Capua's insights, offered to me via email.

57. For "working through" and "acting out," see Dominic LaCapra, *Representing the Holocaust: History, Theory, Trauma* (Ithaca, N.Y.: Cornell University Press, 1994), esp. 204–24.

58. "By denying the reality of others' suffering, people not only diffuse their own responsibility for the suffering," sociologist Jeffrey C. Alexander asserts, "but often project the responsibility for their own suffering on these others" ("Toward a Theory of Cultural Trauma," in Jeffrey C. Alexander et al., eds., *Cultural Trauma and Collective Identity* [Berkeley: University of California Press, 2004], 1–30 [quotation on 1]).

59. Glenda Elizabeth Gilmore, *Gender & Jim Crow: Women and the Politics of White Supremacy in North Carolina, 1896–1920* (Chapel Hill: University of North Carolina Press, 1996), 92.

60. Cathy Caruth, *Unclaimed Experience: Trauma, Narrative, and History* (Baltimore: Johns Hopkins University Press, 1996), 1.

61. Liz Pearsall, "Defending La Porte—Hurricane Ike," *La Porte Examiner,* August 7, 2010, www.examiner.com/article/defending-la-porte-hurricane-ike.

THE 1928 HURRICANE IN FLORIDA AND THE WIDER CARIBBEAN

CHRISTOPHER M. CHURCH

When we think of Florida, the Gulf of Mexico, and the wider Caribbean today, we often think of sandy beaches, retirement homes, vacation cruises, and seaside resorts. In contrast to these blissful thoughts, however, stand the devastating consequences of storms like Hurricane Katrina in 2005 or Hurricane Andrew in 1992. The entire history of the Greater Caribbean Basin—the semi-unified region from the coast of South Florida through the archipelago of the West Indies and along the entire coastline of the Gulf of Mexico—is one of risk-taking in the face of ecological hazards. And yet, despite the inherent dangers that marked the region's rhythm of life, many looked to this region for wealth and, more recently, leisure. It was a place where fortunes were made and lost, where lives were transformed for better or worse. Its coves served as havens for pirates; its reefs sunk Spanish galleons; and its scorching sun menaced the region's countless unfree laborers.

Nothing stands more starkly as a reminder of the inherent risks of life in the Greater Caribbean Basin than the annual arrival of the hurricane season—a perpetual threat to both life and property. This is particularly true when it comes to the hurricane of 1928—a category-five storm with sustained winds over 155 miles per hour that decimated the West Indies, killing over a thousand on the French island of Guadeloupe and hundreds in Puerto Rico, before laying waste to Florida as a category-four cyclone. As the second-deadliest "natural" disaster ever to strike the United States and one of the most prominent storms in history, the 1928 hurricane stands as a painful reminder of the region's inherent ecological risks as well as its colonial legacy and lasting sociopolitical inequities.

The Gulf South can be situated within a larger narrative about the Caribbean Basin as a whole. The early twentieth century was a time of change for the region. Slavery had ended, to be replaced by poorly paid industrial and agricultural wage labor and supplemented by inequitable state practices and institutionalized racism. The so-called American Century had begun, with the United States flexing its muscles in the West Indies and looking to develop the Florida peninsula for economic gain. At the heyday of American expansionism and on the eve of the Great Depression, economic development in Florida contributed to, and was influenced by, the region's unrelenting environmental volatility and economic reconfiguration in an ever-changing global sugar market.

The 1928 hurricane was a regional hazard that produced a devastating disaster when it intersected with ecological terra-forming, short-term economic gains against environmental risk, social and racial inequity, and the development of what sociologist César Ayala calls an "American Sugar Kingdom." While the hurricane had varied consequences across the Greater Caribbean—from failed land speculation and economic gambles in South Florida, to mounting military intervention and economic interest in the West Indian archipelago, to offshore banking, to the short-term collapse of the sugarcane industry—the hurricane can be related specifically to economic development and agricultural practices in the region, interacting with a global sugar economy up for grabs following the devastation of the First World War and the rampant growth of the 1920s. By tracing the hurricane's entire path of destruction and its consequences across the Gulf and Caribbean rather than focusing solely on Florida, we can arrive at a clearer picture of the peculiarities of the region in the 1920s, a greater appreciation of the ecological risks that framed everyday life in the Greater Caribbean Basin, and a better understanding of disasters and their transnational consequences.

Setting the Stage: Hurricanes in the Greater Caribbean Basin

Hurricanes have long been a defining experience for those who inhabit the Greater Caribbean Basin—the geographic region that encompasses the coastline of the Gulf of Mexico along the southern United States and Central America, the Greater and Lesser Antilles, and the north-

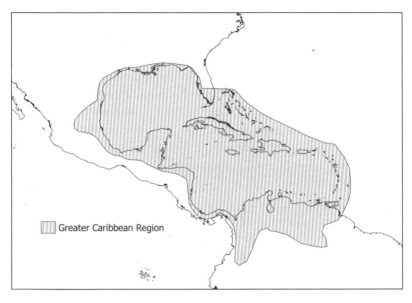

The Greater Caribbean as defined by its hurricane ecology and historical economic development. Map by author, based in part on map in Victor H. Rivera-Monroy, Robert B. Waide, Jorge Herrera-Silveira, and Manuel Maass, "Establishing Hurricane Network in the Greater Caribbean Region," *Long Term Ecological Research Network News* 22, no. 1 (Spring 2009).

ern coast of South America. According to historian Stuart Schwartz, hurricanes provide a leitmotif that helps to explain Caribbean society, because these storms "have been a determining force in the patterns of the region's history."[1] In fact, the first recorded hurricane in the region occurred during Christopher Columbus's second voyage to Hispaniola in 1494, when Isabella became the first European settlement in the New World destroyed by gale-force winds.[2] As one observer put it following the 1928 storm, hurricanes "have scourged the earth from time immemorial," and their arrival in Florida, "a region that has been lashed repeatedly by tropical gales," "is as old as the hills."[3]

The key question, however, is what transformed a gale-force storm— a perennial hazard of Caribbean ecology known to all—into a disaster that shook the United States and the Greater Caribbean during a time of economic prosperity after the First World War. Ecological hazards are not in and of themselves disasters, but become disasters through their intersection with human society: as Lowell J. Carr, the Michigan sociologist who helped found the field of disaster studies, explained, "not every windstorm, earth-tremor, or rush of water is a catastrophe. . . . It is the collapse of cultural protections that constitutes the disaster proper."[4] To put it differently, and to quote Russell Dynes, "natural disasters are social, rather than natural, phenomena."[5] To understand how the well-known, virtually guaranteed arrival of the Atlantic hurricane season transformed an ecological hazard into a societal catastrophe, we now turn to the economic and social circumstances of the 1920s in Florida and the Caribbean that made an ecologically typical cyclone into an aberrant human disaster.

A Recipe for Disaster: Terra-forming, Industrial Agriculture, and Population Growth

Nearly all of South Florida had once been wetland—a marshy mixture of water and sawgrass hospitable to birds, alligators, and fish, but not to humans. Yet, by the late 1920s, this had changed dramatically with a "rush for Florida lands" that was reputed to have "made thousands fabulously rich over night and to have caused cities to rise almost by magic from reclaimed swamp lands or barren wastes of sand."[6] Unfortunately, as the parable goes, cities—like houses—cannot stand when built upon sand, or in this case, on Everglades muck. While the original, prehuman landscape could absorb the arrival of tropical storms and hurricanes, this new anthropogenic landscape spelled disaster, as the dramatic changes in Florida's landscape had set the stage by 1928 for a cataclysmic confrontation with nature.

Beginning in the 1880s, agro-industrialists and the U.S. Army Corps of Engineers began the colossal project of draining the 700,000 acres of wetland south of Lake Okeechobee known as the Everglades. Through a series of canals and locks, dredging, and the importation of invasive species like the melaleuca tree—an Australian species of paperbark able to absorb vast quantities of water—the U.S. business elite and the Army Corps of Engineers radically transformed the ecology of South Florida, and in just under thirty years, capital investment in the Everglades had accomplished the radical ecological change needed for plantation economies—a process that had been taking place in the Caribbean since the seventeenth century. Over the next five decades, the Everglades would serve as the locus of agricultural and commercial development that spurred a population boom in Florida between 1900 and 1920, as the state's burgeoning agro-industry brought itself into competition with the Caribbean by creating what scholar Gail Hollander calls an "imagined economic geography." Seeing Florida as the prototypical "other" to Cuba and the Greater Antilles, Florida's agro-industrialists envisioned "an elaborate economic landscape that would transform a previously undeveloped region" into the "Sugar Bowl" of the United States.[7] To do so, Hollander explains, Florida's boosters used nationalist rhetoric to argue that the United States needed to put its American sugar peninsula to work. According to a broadside put out by one such booster group in Jacksonville, the All-American Sugar League, only "a large American cane-sugar industry" centered in Florida could cure the "American sugar shortage," for a single "cane-field in Florida 45.52 miles square will produce all the sugar consumed in America."[8]

To this end, technocrats, engineers, and the business elite began the process of turning Florida's mucky wetlands into fertile farmland. Following the 1913 Randolph Report, which outlined the master plan for the draining of the Everglades, engineers began cutting canals throughout South Florida. While funding for the infrastructure projects was hard to secure and the Randolph plan was far from completed as late as the 1920s, agricultural investment in South Florida was rapid. It had been discovered that the soil in the Everglades—known locally as "muck"—proved to be incredibly fertile and was, by the 1920s, being excavated and sold as fertilizer.[9] In "establishing a great American cane-sugar and sirup industry" by transforming the Everglades into the "Agro-Industrial Empire of the South,"[10] investors radically changed the ecology of South Florida and dramatically increased the region's vulnerability to environmental hazards. When Governor John Martin called Florida "a pioneer state" built upon "unrivalled agricultural and horticultural possibilities" in 1925,[11] the state was in the process of accomplishing in decades what it had taken the rest of the Caribbean centuries to accomplish: the clear-cutting of natural forests, the draining of wetlands, the importation of invasive species, and the creation of vast cash-crop plantations.

Yet, the goal for interested agro-industrialists was to figure out how to terra-form the land while expending the least amount of capital, and, as historian Ted Steinberg has rightfully argued, to willfully ignore and even downplay the very real hazards posed by Florida's Caribbean ecology in order to spur economic development via investment in real estate.[12] As Wallace R. Moses, a prominent orange grower near Fort Pierce, wrote to Joshua C. Chase, an insurer and seller of agricultural supplies, a month after the 1928 hurricane, "It would be a pity not to utilize this great area of muck lands for such crops as are adapted to it . . . [but] I am inclined to agree with you that spending huge sums of money to build dikes is an extravagance."[13] Faced with the rapid influx of population during the 1920s, the Everglades Drainage District built a forty-seven-mile system of earthen dikes between five and nine feet high along the southern shore of Lake Okeechobee. This system of modest and ultimately inadequate dikes, which broke during the storm and killed approximately two thousand, was the main piece of infrastructure guarding Florida's sugar lands from inundation. For those financiers who stood to profit from the agricultural exploitation of the Everglades, the cost of building the necessary infrastructure to protect the farmworkers' communities was simply too steep. However, the rapid growth of Florida's agro-industry around the lake, and the experience of the 1928 hurricane, would prove such a position to be folly. As the Everglades Drainage Commission explained in 1929, "Experience during the hurricane of 1928, again emphasizes the absolute necessity for protection against Lake Okeechobee," and this protection must be completed at no delay because "records are often broken and there is no way of forecasting how soon another [hurricane] may be experienced."[14]

Compounding the risks concomitant with inadequate storm prevention measures, Florida's population exploded during the early decades of the twentieth century. Outside of the American West, Florida was the fastest growing state between 1900 and 1950.[15] In 1880, the entire state had a population of only 269,000, with the largest city not on the mainland but on the island of Key West off the southern tip of Florida.[16] That put the entire state's population in 1880 on par with that of many of the small Caribbean islands. However, this all changed rapidly due in large part to two massive feats of engineering: the drainage of wetlands south of Lake Okeechobee and the creation of railway lines, most notably the trains reaching into South Florida in the 1890s and the opening of the Flagler line through Miami to Key West in 1912. In the forty years from 1880 to 1920, Florida's population nearly quadrupled from under 300,000 to nearly 1 million.[17] The Greater Miami metropolitan area, incorporated only in 1896, saw its population double between 1900 and 1910, and then quadruple between 1910 and 1920. By 1930, South Florida's population alone had grown to over 250,000, with the vast majority living in Dade, Palm Beach, and Broward counties.[18]

Beginning in 1921, Florida's east and west coasts began to blossom, as northern visitors—so-called "Tin Can Tourists," named for their packing of tin cans for their stays at Florida resort camps—purchased vacation homes in the Sunshine State.[19] As the 1920s roared on, rampant land speculation and short-term profit-seeking skyrocketed, and Florida's population continued to climb. By the end of the 1920s, the state's population had grown to nearly 1.5 million,[20] and construction had begun to strip Florida of many of its natural protections from hurricanes: mangrove trees whose roots anchor soil, coastal berms that provide protective walls, barrier islands that slow gale-force winds and harbor the shore, and natural wetlands that absorb storm surges and evenly distribute copious amounts of precipitation.[21]

However, while Florida's population exploded on the whole during the first two decades of the twentieth century, it did so unevenly. Many settled along Florida's coastline, but only the most intrepid or desperate of Florida's recent immigrants settled in its swampy hinterland. By the start of the 1920s, the bulk of the Everglades remained undrained and typically flooded from September to February, and Florida's inland areas even witnessed an outmigration in the early years of the Roaring Twenties.[22] With real estate advertising that claimed agriculture in Florida promised "big money" and "freedom from wage slavery," this would soon change, however.[23]

South Florida's first real estate bubble began in 1924 and peaked in 1925, with land prices rising 200 to 300 percent in some areas.[24] The rapid land speculation by faraway investors lasted until 1926, when a category-four hurricane decimated the Greater Miami area and put an end to Florida's short-lived economic boom. Normalized to 2006 dollars, the 1926 Miami Hurricane caused the most economic damage in the past century, with somewhere between $140 and $160 billion worth of damage. By contrast, Hurricane Katrina in 2005 caused $81 billion, and the deadliest hurricane in U.S. history, the 1900 Galveston hurricane, caused $99 billion of damage.[25] The 1926 hurricane brought land speculation in Florida to a grinding halt, as land prices in some locales dropped to one-hundredth of their 1925 values. Land that had once sold for $60,000 could now be purchased for a mere $600.[26] The rapid development in South Florida had proved to be a dual-edged sword, as population density and urban development compounded the disastrous effects of the hurricane. For instance, if a storm like that of 1926 were to hit South Florida in 2020, it has been estimated that the amount of damage would be over $500 billion—approximately four times more costly.[27]

As real estate prices plummeted, a renewed focus was placed on draining the Everglades for development, with Florida's legislature supporting drainage schemes via a system of taxes and bonds. Backed by Florida's governor, John Martin, the plan would deepen existing canals and create a complex system of locks and dams to further lower the water table and drain the Everglades to create more arable land. At first, prominent businessmen were against the bond scheme on the grounds that the government was unnecessarily interfering in the market. Yet their focus on the possibility of profits assuaged their concerns over government intervention and blinded them to the ecological dangers, putting the state further at risk for the devastation that was to come in 1928. As contemporary Joshua Chase explained in 1927, "the state of Florida would be pledging its money and credit to buck up a lot of real estate speculation schemes" centered on draining the Everglades, and the state's speculation bonds "now in the hands of people outside of the state, bought at a discount, would immediately become awfully good [financial] security."[28]

At the same time that South Florida was being terra-formed via drainage canals and dikes, as well as interconnected via railway and road construction, it was also experiencing the massive growth of a sugarcane industry, artificially propped up via subsidies to compete with Caribbean sugar. Between 1890 and 1930, per-capita sugar consumption in the United States had nearly doubled as Americans honed their sweet tooth and flexed their muscle in the Spanish Caribbean.[29] This mounting demand meant that each American consumed nearly 110 pounds of sugar per year by 1930.[30] While the vast majority of this sugar was imported, mostly from U.S. interests in Puerto Rico, Cuba, and the Dominican Republic, astute investors saw the potential for growth and capital gains in South Florida's fledgling domestic sugar industry.

As the state's agricultural land in the Everglades became more developed, dominated by the production of winter vegetables until midcentury, large-scale agro-business began constructing employee villages to accompany its various plantations, seeking to house migrant labor during the planting and harvesting seasons. These "plantation villages" were modeled on the older slave shanties of the proto-industrial plantations during the nineteenth century—only now they housed poorly paid migrant workers rather than unpaid laborers. Even the language used to describe Florida's budding sugar production and blossoming winter vegetable farms in the Everglades was full of older plantation language, and the system was much the same—with a plantation overseer supervising a workforce that inhabited company villages, shopped at company stores, grew subsistence crops on company-leased land, and by and large lived at the company's whim.[31] The small villages were self-sustaining agro-industrial towns modeled on the nineteenth-century plantation to the point that, on the day of the annual sugar harvest festival, "Everyone from the plantation overseer to the youngest child on the plantation prays for good weather."[32]

Over the course of the 1920s, Florida was positioning itself to become the United States' sugar colony of the North Caribbean. With the United States controlling Cuba and Puerto Rico after the Spanish-American War, as well as the Panama Canal following its completion in 1914, U.S. officials and the business elite were interested in developing Florida as a strategic holding in the Greater Caribbean, with Miami serving as its chief city, the agricultural land around Lake Okeechobee as its heartland, and Key West its chief naval port. In the words of a 1921 bulletin from the Florida Department of Agriculture, Key West was "The Gateway to the Panama Canal" and the "natural gateway to Caribbean and South American ports."[33] And in many ways, Florida shared much in common with the region whose economic development set the precedent for the state's recent influx of capital investment. Just as financiers and government policies helped to create an "American Sugar Kingdom" in the Caribbean between 1900 and 1930, capital investment, wetland draining, and African American migration would transform Florida into a sugar-producing behemoth by the middle of the century— one plagued by all the characteristics of an exploitative colony.

This development had profound effects on Florida's demographics. While Florida's population was nearly split equally between white and black in 1900, by 1930 recent white immigrants outnumbered black residents almost 2.5 to 1.[34] However, the mass influx of white immigrants settled mostly along the coast, while the population in the agricultural area around Lake Okeechobee was still majority black.[35] While only 30 percent of Palm Beach County's population of 51,000 in 1930 was black, virtually all African Americans and Afro-Caribbeans in the county lived along the shores of Lake Okeechobee. The town of Pahokee on the shore of Lake Okeechobee, which has historically been predominantly African American and tied to the sugarcane plantations, was 56 percent black. In 1930, roughly 1,500 of its 2,700 residents identified themselves as "negro" on the 1930 census, and most of these were field laborers from the American South who migrated for work and Bahamians brought in for the winter vegetable and sugar harvest. The migrant farmhands either lived in plantation villages or rented homes for as little as one dollar per month.

As with the new agricultural landscape in South Florida, the Caribbean islands had been historically dependent upon a variety of monocultures, foremost among them sugar. Sugarcane ravages the environment by depleting the soil, and it requires large-scale terra-forming activities—such as deforestation, leveling, draining, and water diversion—to transform tropical landscapes into fertile agricultural land. As part of the resurgence of tropical cane sugar in global markets, Florida's population was booming and its fledgling sugar economy growing. Likewise, the Caribbean islands, which during the late nineteenth century were having an increasingly difficult time finding their place in an international sugar economy dominated by European beet sugar, were seeing their sugar economies resurge as sugarcane grown in the tropics once again supplanted beet sugar by 1930. However, this resurgence was disproportionately felt in the Spanish-speaking Caribbean, where, as Ayala puts it, "U.S. sugar refining industrialists built an empire of sugar plantations in the Caribbean after 1898" via tariff manipulation, vertical integration, and aggressive business tactics.[36] Those parts of the Caribbean outside of the American sphere of influence—such as the French Caribbean, intimately tied to metropolitan France as it was—were left at a substantial disadvantage and saw sugar production continue to decline in the long term.

Cutting a Path through the Antilles

It is against this backdrop of economic investment and ecological risk-taking in South Florida, and economic resurgence in the Caribbean, that we now turn to the 1928 hurricane. The typical path for an Atlantic hurricane takes it first across the Lesser and Greater Antilles, subsequently through the Bahamas, and then either along the eastern Florida coastline into the Carolinas or across Florida into the Gulf of Mexico. The latter was the case with the 1928 hurricane, which cut across the state of Florida after leaving a path of destruction across the Antilles.

On September 12, 1928, the storm—now a category-four hurricane—made landfall on the French island of Guadeloupe, a small bastion of nonwhite French civilization and the once-predominant producer of colonial sugar in the Lesser Antilles. The hurricane caused massive damage to the island's capital of Basse-Terre and spawned a tornado that ravaged its largest city of Pointe-à-Pitre. The initial death toll was estimated northward of 800, and would later top 1,200. With the devastation on the island "immense," the governor of Guadeloupe telegraphed the neighboring island of Martinique to request supplies. In addition to fuel, cotton, mercury, and iodine, the governor—fearful of civil unrest—requested "as many soldiers as you could spare."[37] However, while Martinique sent the requested supplies, it had no troops to send to Guadeloupe, as all supplemental military personnel had been dispatched in August to French Guyane to deal with the civil unrest in Cayenne that erupted after the death of Jean Galmot.[38] While the hurricane of 1928 caused very few casualties on Martinique, it nevertheless caused a great deal of property damage, prompting a substantial number of claims to the French government for state aid in the form of direct compensation and/or tax exemption.[39] As shown by Caribbean economic historian Christian Schnakenbourg, the Okeechobee hurricane and its devastation of the 1928 sugar crop plunged the French Caribbean into an economic spiral that worsened with the Great Depression of the 1930s.[40]

Two days later, on September 14, the storm passed into the United States' sphere of influence by way of the Virgin Islands, before making landfall in Puerto Rico—the United States' key colonial holding in the Caribbean. This was not the first devastating storm to threaten U.S. interests on the island. Less than a year after the end of the Spanish-American War, which had passed possession of the island from Spain to the United States, while the colony was still under military occupation, a category-four hurricane slammed into Puerto Rico and killed over three thousand people. As the deadliest storm in Puerto Rico's history, the 1899 San Ciriaco hurricane disrupted the island's economy so badly that it had not yet recovered when it was struck again nearly thirty years later in 1928.

Much like its predecessor in 1899, the 1928 storm, named after San Felipe in Puerto Rico, was devastating. It killed as many as one thousand people, hitting the island's poor rural population the hardest and leaving up to one million homeless during the aftermath.[41] While the initial impact was not as great as with Guadeloupe, the disease epidemic that followed in its wake crippled the small island, prompting the alarmed governor, Horace Towner, to telegraph Washington for help.[42] Cases of malaria nearly tripled during the storm's aftermath, while influenza saw nearly a 4,000 percent increase. In 1927, there were a mere 136 cases of influenza, but in 1928 the number skyrocketed to 5,379. The American Red Cross used over 2.8 million units of diphtheria antitoxin, 113,980 vials of typhoid vaccine, and nearly 4.2 million units of tetanus vaccine to combat the widespread epidemics.[43] Within days of the disaster, temporary hospitals were set up in the hardest-hit areas, and the Florida press warned, "Epidemic adds to storm woes in Porto Rico [sic]."[44] The initial destruction and following disease served as a warning to nearby Florida, where conditions were also ripe for a disease outbreak in the rural area around Lake Okeechobee, and it seemed to belie the notion, put forth by the American Red Cross, that Puerto Rico's climate was "altogether favorable to the Caucasian race."[45]

According to contemporary reports, most important was not the storm's effects on urban and rural dwellings, nor the outbreak of disease, but the hurricane's crushing effect on the island's monoculture economy. Crop losses were placed at $46 million, with the overall agricultural losses (both crop and equipment) totaling $69 million. This preoccupation with sugar cultivation was a common thread that weaved through Guadeloupe, the American sugar colony of Puerto Rico, and Florida, the newest site of sugar plantations in the mainland United States.[46] Due to U.S. investment, sugar production in Cuba, Puerto Rico, and the Dominican Republic increased by nearly 1,600 percent between 1900 and 1925,[47] radically altering the economic life of the Caribbean. After American acquisition of Puerto Rico, the United States had pushed the Puerto Rican economy from coffee to sugar cultivation. Whereas in the late nineteenth century, Puerto Rico produced four times as much coffee as sugar, by the 1920s Puerto Rico's primary export was sugar—a fact solidified by the devastation of Puerto Rico's coffee crop during the 1928 hurricane.

All of these "sugar outposts"—from Florida to the Virgin Islands and Puerto Rico—were built upon the backs of African laborers. That is, the laboring classes of the Caribbean were primarily of African descent, with the islands' capital centralized in the hands of a minority of European, and increasingly American, landowners. Therefore, initial reports of the storm were seen through the lens of American colonial ambitions in the region, and menaced by the fear of colonial danger. As the storm prepared to strike South Florida, observers in Miami were in awe of the damage in Puerto Rico, which was worse than they had initially thought, as well as relieved that "no continental American lost his life in the storm, the death list being confined to natives."[48]

The Storm Hits Florida

Leading up to the hurricane's landfall in Florida, U.S. officials were convinced that the so-called "West Indian Storm" would miss Florida and consequently posed "no danger of winds of greater than gale force for Miami and the extreme lower east coast of Florida."[49] While the storm had affected more than 60 percent of the island populations in Puerto Rico, the Virgin Islands, Guadeloupe, and the Bahamas,[50] the American press believed—or rather hoped—that the hurricane would be a phenomenon isolated to those unfortunate islands south and east of the Florida peninsula. However, they would be proved wrong on the afternoon of September 16, when the hurricane made landfall at West Palm Beach, before progressing up the center of the state the following day. The storm, which had unleashed its fury on Guadeloupe five days before, had now brought its destruction to the continental United States, causing substantial wind damage along Florida's coastline and massive amounts of flooding in the wetlands and sugar plantations around Lake Okeechobee, the state's largest freshwater body. The worst of the storm did not hit Florida's coastal towns, which had been well prepared for the coming of another storm in the wake of the Miami hurricane of 1926. Weather bulletins and news reports had warned coastal residents to shelter in place, and the homes that were destroyed were "light timber houses and poorly built houses."[51] The worst hit the denizens of the center of the state, who had been the poorest, least prepared, and most vulnerable.

With the inevitable landfall of the storm, many remembered the

1926 hurricane two years prior and had evacuated the towns of the western shore of Lake Okeechobee. Yet, the same was not true of the burgeoning agricultural towns along the southern shore. As the storm pushed inland, it ruptured the dike at Belle Glade, which kept Lake Okeechobee's waters from flowing into the sprawling Everglades. The Everglades—or River of Grass—had been drained to create the vast sugar plantations and agricultural fields that had drawn so many itinerant workers to South Florida, and with the breaking of the dike, water returned to its normal southward flow with a vengeance, creating twenty-foot-high waves that spread out thirty miles in all directions. The raging waters rose to eight feet at a rate of one inch per minute, engulfing the small laborers' town of Belle Glade and completely inundating the former wetlands—now a "region of compressed debris and stench." This not only destroyed the recently drained farmland, but drowned 1,770 agricultural workers and left roughly fifteen thousand families homeless, with six thousand families left destitute in Palm Beach County alone.[52]

As social occurrences, disasters unfold unevenly along socioeconomic divides, and circumstances in the Everglades mirrored those of the West Indies. Tensions were high between the recent influx of white middle-class families who settled in the coastal tourist towns of Boca Raton and West Palm Beach, and migrant workers of color and Florida's poor white population, who worked the cane and vegetable fields and lived in neo-plantations in the center of the state around Lake Okeechobee—"a thriving farm center devastated in its embryonic stage" where its agricultural water supply at Lake Okeechobee was now "a muddy, brownish fluid, too heavy and smelling with hundreds of corpses of men and beasts on its breeze-tossed surface to be called water."[53] This discord manifested itself in press coverage of the disaster. Reports distinguished between numbers of "negro" and white casualties, alarmed far more by reports of the second than of the first. With reports of "native" deaths in the Caribbean, particularly in the American colony of Puerto Rico, observers in Florida were relieved that this "West Indian" storm had yet to affect "real" American lives. As the storm made landfall in South Florida, this weighing of "American" lives against those of "others" continued as the storm blazed across Lake Okeechobee. It was clear to any observer that African Americans in disaster-stricken South Florida were seen as toilers and laborers, rather than as true citizens of Florida. As relief organizers put it during the immediate aftermath of the storm, "a large force of negroes is being organized in the stricken area to help clear the streets and search the wreckage."[54]

The agricultural devastation around Pahokee and Lake Okeechobee affected, in the words of observers, "scores of negroes" who worked on the sugar farms there.[55] According to the U.S. Department of Commerce's compendium of mortality statistics, "there were at least 1,524 deaths (mostly of colored persons) due to the hurricane," as reported by the Florida State Board of Health.[56] In fact, the bodies recovered from the farmlands just south of Lake Okeechobee were sorted by skin color and treated differently. In one case, recovered corpses were buried in separate cemeteries in West Palm Beach, with 68 white bodies buried in Woodlawn Cemetery and 674 "colored" buried at City Cemetery.[57] In another case, 55 white bodies were buried, while another 212 "negroes" were disposed of unceremoniously, with few receiving a burial and most others being burned in a mass funeral pyre.[58] Roughly 1,500 people were buried in a mass grave at Mayaca Cemetery outside of Pahokee. Officials found it simpler to burn the bodies of African Americans and Afro-Caribbeans on the spot or to bury them in mass graves than to transport them to the nearest cemetery.

The racial divisions were intricately tied to the agro-industrial sugar and vegetable plantations, and were a direct consequence of the quick reconfiguration of the ecological and social landscape of Florida in the early decades of the twentieth century. The town of Pahokee, which had seemingly sprouted overnight, was a predominantly black laborers' town on the edge of Lake Okeechobee. It had been hit particularly hard by the storm, mostly affecting the 56 percent of the town's population that was black. Given its lopsided class composition along racial lines—with African Americans comprising the large laboring population and whites the managerial—the press reported that, while only twenty whites were drowned, over three hundred "colored" had been lost to the floodwaters.

There was a particular level of insensitivity to the plight of these recent immigrants to Florida. According to Joshua Chase, the devastation to Florida's sugar settlements was inevitable, for "Lake Okeechobee is a huge saucer," and "If you pour coffee into a saucer, you could easily blow it all out if you blew hard enough." His conclusion, given the vast expense of creating a dike capable of withstanding a hurricane such as that in 1928, was to "let the waters of Lake Okeechobee spread over the surrounding counties just as they have for centuries" and compel those settlers "who insist upon remaining" to "live in house boats at their own risk."[59] Given that such "settlers" had little choice, due to their socioeconomic status and the need to work the fields for their livelihood, the sentiment betrayed the exploitation inherent to Florida's emerging agro-industrial colony along Lake Okeechobee.

Aftermath

The 1928 hurricane caused between $31 and $34 billion worth of damage in Florida alone, normalized to 2006 dollars.[60] In all, the Red Cross at the time estimated that 1,836 people perished in Florida, most of whom had been migrant farmworkers along Lake Okeechobee's southern shore. However, it is highly likely that this figure underreports the number of casualties. Many bodies had been swept away by the storm waters, never to be recovered, and Florida's growing tourism industry had every reason to minimize the number of deaths to prevent potential visitors from seeing only Florida's ecological hazards and not its economic and leisurely opportunities. This did not work out as planned, however, for as historian Mike Davis has pointed out, the 1926 Miami and 1928 Okeechobee hurricanes disadvantaged Florida in the race to win tourist dollars and in agricultural development for an entire generation.[61]

Nevertheless, the 1926 and 1928 hurricanes prompted the U.S. Army Corps of Engineers to institute new flood controls and to buttress the massive dike-and-levee system with the Herbert Hoover Dike in 1930. This dike diverts overflow from Lake Okeechobee into a series of canals that ultimately empty into the Atlantic Ocean, keeping dry the thousands of acres of farmland on the south shore. By the 1930s, Florida was a hodgepodge of various agricultural endeavors, from citrus to tropical fruits to cattle to sugarcane. Before the storm, sugarcane cultivation was mostly limited to Clewiston on Lake Okeechobee's west shore and the rural areas around Belle Glade on the southern shore. Despite the devastation, the storm did little to slow Florida's sugar development, much to the detriment of the Caribbean outside of U.S. control, namely the French Antilles, which found it increasingly difficult to compete with American-dominated sugar interests. While only about 7,000 acres of sugarcane were under cultivation in Florida by the end of the 1920s,[62] this acreage had doubled by the 1930s and witnessed a substantial takeoff in the late 1940s and early 1950s and then again after the Cuban embargo in 1961, supplanting Florida's historically more dominant industry of winter vegetable cultivation. Today the majority of the 700,000 acres of agricultural land in the Everglades is devoted to sugarcane cultivation, with roughly three-quarters of all sugarcane fields located in Palm Beach County alone.[63]

The United States had been flexing its economic and political muscle in the Caribbean Basin at the same time that it was developing the Florida peninsula as its gateway to the tropics. The Virgin Islands had been part of the United States for only a little over a decade before the arrival of the 1928 hurricane. Purchased from Denmark in 1916, the Virgin Islands represented the latest in a long line of territorial acquisitions and a growing sphere of influence in the Caribbean—from the annexation of Puerto Rico in 1898, to the submission of Cuba via the Platt Amendment in 1901, to the dual military occupations of Haiti from 1914 to 1935 and the Dominican Republic from 1916 to 1924. If anything, the United States redoubled its efforts in creating and maintaining Florida as the "gateway to Panama" and the capital of the Caribbean following the 1928 hurricane.

The 1928 hurricane slowed Puerto Rico's meteoric rise in sugar production, but did not stop it. Although the crop that year was decimated, and the 1929 harvest was lower than it had been in previous years, sugar production on the island was once again on the rise by 1930.[64] Coupled with the 1926 hurricane, the 1928 storm crippled the production of coffee, the once predominant staple of Puerto Rico, and vastly favored the expansion of the Puerto Rican sugar economy, which was already propped up by favorable U.S. tariffs.

U.S. investment did little to buttress those islands outside of the country's sphere of influence, however. Guadeloupe's role as one of the world's leading sugar producers had been in decline since the world-wide sugar crisis of the 1890s. Although the island witnessed a rebound in sugar production during the first decades of the twentieth century, particularly due to the First World War's devastating effects on beet cultivation in Europe, by the 1920s production was a fraction of what it had been in the nineteenth century.[65] The 1928 storm destroyed virtually all of that year's sugarcane harvest, in many ways putting the final nail in the coffin of Guadeloupe's sugar industry. While the industry rebounded somewhat during the mid-1930s, dry weather, the world-wide depression, and the lasting effects of the 1928 storm kept sugar production in Guadeloupe in a state of general crisis. As in Puerto Rico, the hurricane destroyed the island's coffee and cocoa industries, which never fully recovered.[66]

Due to the reconfiguring of the world sugar market, including beet-sugar production in the United States and the creation of the American sugar empire in the Caribbean, as well as favorable protectionist measures passed by the French government, Martinique and Guadeloupe both invested heavily in banana cultivation following the 1928 hurricane. While Guadeloupe had grown only 1,431 tons of bananas in 1926, it was growing more than 50,000 tons a decade later. Over that same period, Martinique's banana production grew from a virtually nonexistent industry to one that exported 34,000 tons of bananas to France alone.[67] Following the 1928 storm, Guadeloupe lost any hope of dominating the Greater Caribbean sugar market. Likewise, Guadeloupe's neighbor, Montserrat, lost its already declining sugar economy to the hurricane, with sugarcane cultivation essentially negligible after 1928.[68] Despite the damage caused by the 1928 hurricane, or rather because of it, the establishment of a veritable U.S. sugar empire in the Greater Caribbean seemed like a foregone conclusion.

Conclusion

In 1928, South Florida stood on the edge of a U.S. sugar empire stretching across the Caribbean Basin and increasingly centered in the muck-lands of the Everglades. As migrant workers came to harvest the cane, as they had done in the Caribbean for centuries, they were housed in plantation villages along Lake Okeechobee's southern shore with naught but a nine-foot-tall dike and a handful of canals to protect them. Hurricanes are a defining characteristic of the Greater Caribbean, one intimately tied to the region's geographic and economic history. When the so-called West Indian storm of 1928, which failed to claim any "true" American lives during its inexorable march across the Antilles, landed in South Florida, it stood alongside the state's plantation economy as a bitter reminder that the Gulf South belongs to the Caribbean. Evidencing the inherent risks of the Greater Caribbean, the hurricane intersected with recent economic and demographic development in Florida—a state being carved up by canals, dammed with dikes, and covered in railways as tourists settled along the coasts and laborers in the "heartland"—to claim the lives of as many as two thousand African American and Afro-Caribbean laborers, over whose graves would sprout one of Florida's largest agricultural industries: sugarcane.

NOTES

1. Stuart Schwartz, "Hurricanes and the Shaping of Circum-Caribbean Studies," *Florida Historical Quarterly* 83, no. 4 (2005). 305–409.

2. A. Romer, *Les Cyclones à la Martinique* (Fort-de-France, Martinique: Imprimerie du Gouvernement, 1932), 6.

3. C. F. Talman, "Tropical Hurricanes Are a World Scourge," *New York Times,* September 23, 1928.

4. Lowell Juillard Carr, "Disasters and the Sequence-Pattern Concept of Change," *American Journal of Sociology* 38, no. 2 (1932): 207–18.

5. Russell R. Dynes, "Disaster Reduction: The Importance of Adequate Assumptions about Social Organization," *Sociological Spectrum* 13 (1993): 175–92.

6. "Florida Is Facing Serious Conditions," *New York Times,* January 3, 1927.

7. Gail M. Hollander, *Raising Cane in the 'Glades: The Global Sugar Trade and the Transformation of Florida* (Chicago: University of Chicago Press, 2008), 15–17.

8. "Sugar, Produce It in America with American Labor and Capital," broadside (Jacksonville: Drew Press, 1918), floridamemory.com/items/show/212502 (accessed November 18, 2014).

9. "Moore Haven Muck Shipped by the Carload for Fertilizer," broadside (DeFunlak Springs: Catts' Realty Company, 1922), floridamemory.com/items/show/212448 (accessed November 18, 2014).

10. *The Everglades: Agro-Industrial Empire of the South* (Clewiston, Fla.: United States Sugar Corporation, 1944).

11. Frank Parker Stockbridge and John Holliday Perry, *Florida in the Making* (New York: de Bower Publishing, 1926).

12. Ted Steinberg, *Acts of God: The Unnatural History of Natural Disaster in America* (New York: Oxford University Press, 2000), 47–75.

13. Wallace R. Moses, "Correspondence between Wallace R. Moses and Joshua Chase regarding Lake Okeechobee and Hurricane Season," Business Records of Sydney Octavius Chase (1928), Box 22, Folder 5.68, Chase Collection, Special and Area Studies Collections, George A. Smathers Libraries, University of Florida, Gainesville.

14. F. C. Elliot, *Everglades Drainage District Biennial Report 1927–1928 to the Board of Commissioners of Everglades Drainage District* (Tallahassee: T. J. Appleyard, 1929), ufdc .ufl.edu/UF00075731/00002 (accessed November 11, 2014).

15. Frank Hobbs and Nicole Stoops, *Demographic Trends in the 20th Century: Census 2000 Reports (CENSR-4)* (Washington, D.C.: U.S. Census Bureau, 2002), 27, www.census .gov/prod/2002pubs/censr-4.pdf (accessed August 28, 2014).

16. Carlton J. Corliss, "Henry M. Flagler: Railroad Builder," *Florida Historical Quarterly* 38, no. 3 (1960): 195–205.

17. U.S. Census Bureau, *1921 Census Report,* www.census.gov/history/pdf/1921_Annual.pdf (accessed August 28, 2014).

18. U.S. Census Bureau, *1930 United States Census,* vol. 3: *Reports by States, showing the composition and characteristics of the population for counties, cities, and townships or other minor civil divisions,* pt. 1: *Alabama-Missouri* (Washington, D.C.: GPO, 1932), www2.cen sus.gov/prod2/decennial/documents/10612963v3p1.zip (accessed November 1, 2014).

19. Jay Barnes, *Florida's Hurricane History* (Chapel Hill: University of North Carolina, 1998), 103.

20. U.S. Census Bureau, *1931 Census Report,* www2.census.gov/prod2/statcomp/docu ments/1931-02.pdf (accessed August 28, 2014).

21. Keryn B. Gedan, Matthew L. Kirwan, Eric Wolanski, Edward B. Barbier, and Brian R. Silliman, "The Present and Future Role of Coastal Wetland Vegetation in Protecting Shorelines: Answering Recent Challenges to the Paradigm," *Climatic Change* 106, no. 1 (2011): 7–29; Edward B. Barbier, "Natural Barriers to Natural Disasters: Replanting Mangroves after the Tsunami," *Frontiers in Ecology and the Environment* 4, no. 3 (2006): 124–31; R. Costanza, O. Pérez-Maqueo, M. L. Martinez, P. Sutton, S. J. Anderson, and K. Mulder, "The Value of Coastal Wetlands for Hurricane Protection," *Ambio* 37, no. 4 (2008): 241–48.

22. Michael Grunwald, *The Swamp: The Everglades, Florida, and the Politics of Paradise* (New York: Simon & Schuster, 2007), 182.

23. "Make Big Money Growing Papayas[:] Win Freedom from Wage Slavery," broadside (Sebring, 1943), floridamemory.com/items/show/212500 (accessed November 13, 2014).

24. Barnes, *Florida's Hurricane History*, 126.

25. Roger A. Pielke Jr., Joel Gratz, Christopher W. Landse, Douglas Collins, Mark A. Saunders, and Rade Musulin, "Normalized Hurricane Damages in the United States, 1900–2005," *Natural Hazards Review*, 2008, www.nhc.noaa.gov/pdf/NormalizedHurricane2008.pdf (accessed August 28, 2014).

26. Barnes, *Florida's Hurricane History*, 126.

27. Pielke et al., "Normalized Hurricane Damages in the United States."

28. Joshua Chase, "Correspondence on February 11, 1927 between Joshua Chase and Sydney Chase regarding Everglades Drainage Bond Scheme," Business Records of Sydney Octavius Chase, Box 22, Folder 5.68.

29. César J. Ayala, *American Sugar Kingdom: The Plantation Economy of the Spanish Caribbean, 1898–1934* (Chapel Hill: University of North Carolina Press, 1999), 30.

30. Ibid.

31. *Sugar and the Everglades* (Clewiston, Fla.: United States Sugar Corporation, 1939), babel.hathitrust.org/cgi/pt?id=c00.31924000871057 (accessed November 18, 2014).

32. Ibid.

33. W. A. McRae, Florida Department of Agriculture, *Florida Quarterly Bulletin* (Tallahassee: Appleyard), vol. 31, no. 1 (1921).

34. U.S. Census Bureau, *1931 Census Report*, www2.census.gov/prod2/statcomp/documents/1931–02.pdf (accessed August 28, 2014).

35. U.S. Census Bureau, *1930 United States Census*, vol. 3: *Reports by States*.

36. Ayala, *American Sugar Kingdom*, 19–28.

37. Archives départementales de la Martinique, FM 1M10318, Fort-de-France, Martinique.

38. Ibid. Galmot had been disgraced in 1919 amid allegations that he had manipulated the rum market, and in 1928 decided to restart his electoral campaign in Cayenne. He died of arsenic poisoning, which prompted his followers to begin a minor insurrection that left numerous dead. His life and death inspired a 1930 novel by Blaise Cendrars titled *Rhum: L'aventure de Jean Galmot*.

39. Archives départementales de la Martinique, FM 1M9891.

40. Christian Schnakenbourg, *Histoire de l'industrie sucrière en Guadeloupe aux XIXe et XXe siècles* (Paris: Harmattan, 2008), vol. 3: 153.

41. "1000 Are Believed Dead in Porto Rico; Property Loss Great," *St. Petersburg Times*, September 17, 1928.

42. "Porto Rican Governor Wires U.S. to Send Help," *St. Petersburg Times*, September 17, 1928.

43. American National Red Cross, *The West Indies Hurricane Disaster, September,*

1928: Official Report of Relief Work in Porto Rico, the Virgin Islands, and Florida (Washington, D.C.: American National Red Cross, 1929), 18.

44. "Epidemic Adds to Storm Woes in Porto Rico," *Miami Daily News*, September 21, 1928.

45. American National Red Cross, *The West Indies Hurricane Disaster*, 10.

46. Ibid., 22.

47. Ayala, *American Sugar Kingdom*, 5–6.

48. "1,000 Persons Believed Dead in Porto Rico," *Miami Daily News*, September 17, 1928.

49. "Winds of Only Gale Velocity Expected Here," *Miami Daily News*, September 16, 1928.

50. American National Red Cross, *The West Indies Hurricane Disaster*, 49.

51. "Storm Effects Told in Survey Made by Radio," *St. Petersburg Times*, September 18, 1928.

52. *Palm Beach Hurricane, Sept. 16, 1928, 92 views*, ufdc.ufl.edu/UF00001306/00001 (accessed November 2, 2014).

53. "Florida Deaths Mounting, Now 800, Many Are Missing; Disease Imperils Living," *New York Times*, September 21, 1928.

54. "2 Deaths, 50 Injured in Palm Beach," *Miami Daily News*, September 17, 1928.

55. "Florida Deaths Mounting, Now 800, Many Are Missing."

56. U.S. Department of Commerce, *Mortality Statistics 1928: Twenty-Ninth Annual Report* (Washington, D.C.: Government Printing Office, 1930).

57. *Palm Beach Hurricane, Sept. 16, 1928*.

58. Ibid.

59. Joshua Chase, October 1928, "Correspondence between Wallace R. Moses and Joshua Chase regarding Lake Okeechobee and Hurricane Season."

60. Pielke et al., "Normalized Hurricane Damages in the United States."

61. Mike Davis, *Ecology of Fear: Los Angeles and the Imagination of Disaster* (New York: Metropolitan Books, 1998), 39.

62. G. H. Synder and J. M. Davidson, "Everglades Agriculture: Past, Present, and Future," in *Everglades: The Ecosystem and Its Restoration*, ed. Steve Davis and John C. Ogden (Boca Raton, Fla.: CRC Press, 1994): 85–116.

63. L. E. Baucum and R. W. Rice, *An Overview of Florida Sugarcane* (Gainesville: University of Florida, IFAS Extension, 2006).

64. James Dietz, *Economic History of Puerto Rico: Institutional Change and Capitalist Development* (Princeton, N.J.: Princeton University Press, 1987), 105.

65. Schnakenbourg, *Histoire de l'industrie sucrière en Guadeloupe*.

66. Ibid.

67. Alain Blérald, *Histoire économique de la Guadeloupe et de la Martinique: Du XVIIe siècle à nos jours* (Paris: Karthala, 1987), 70.

68. *The Generation and Transfer of Technology to Support Cotton Production in the Caribbean*, vol. 3: *Macroeconomic and Agricultural Sector Data of Participating Countries* (San José, Costa Rica: Inter-American Institute for Cooperation of Agriculture, 1992), 117.

SWAMP THINGS

Invasive Species as Environmental Disaster in the Gulf South

ABRAHAM H. GIBSON *and* CINDY ERMUS

People have sought to explain the apparently capricious origins of disastrous events for thousands of years. For most of recorded history, people perceived disasters as acts of God, believing that every calamity, however large or small, was actively prescribed and deliberately executed by one or many supernatural deities. This explanation has fallen out of favor as cultures around the world have grown increasingly science-based, eschewing religious explanations in favor of natural ones. Many insist that disasters are caused by "forces of nature," described as the accidental and, ultimately, random convergence of Earth's various hydrological, climatological, and geological processes. Still others insist that humans are ultimately to blame for disastrous events. According to this view, naturally occurring phenomena like hurricanes and earthquakes merely throw light on already disastrous human processes. This means that people can no longer shift blame for disasters to God or to nature. It is human action and, not infrequently, human *in*action that render any given event disastrous. We have only ourselves to blame.[1]

Most of the scholars who study the history of disasters focus on nonbiological events. Some focus on naturally occurring phenomena, like fires, floods, and tornadoes, while others focus on more clearly human-caused disasters, like nuclear meltdowns, worksite explosions, and toxic spills. In both cases, the catalysts are abiotic in nature.[2] There are important exceptions, however. Many researchers (including two contributors in this volume) study disease-causing microbes within the context of disaster studies, and there is a large body of scholarship devoted to the complex relationship between plagues and peoples. Even so, epidemics remain outnumbered by abiotic disasters. Consider, for example, that one recent encyclopedia lists more than two hundred of the most famous disasters in human history, yet only fourteen of these disasters are biological in nature. Thirteen describe epidemics, and one describes a locust plague.[3]

We sense that historians of disaster might be missing an opportunity. In this essay, we seek to advance the literature on disaster studies by asking whether macroscopic organisms ever deserve to be called disasters. More specifically, we will ask whether introduced non-native populations, sometimes known as "invasive species," qualify as disasters in the traditional sense of the word. We are not the first to do so. In 2011, biologist Anthony Ricciardi and coauthors stated that "biological invasions are fundamentally analogous to natural disasters," and they insisted that we should manage them accordingly. As they explain, treating invasive species as disasters would allow biologists, wildlife management officials, and all levels of government to develop the same kinds of safety codes, emergency preparedness standards, and rapid response measures that are normally reserved for nonbiological disasters. In other words, it would allow biologists and policy makers to implement more precise hazard-reduction plans. "Just as building codes are designed to protect people and structures from earthquakes," the authors write, "we argue that a precautionary system should be in place to manage vectors and pathways to safeguard against all potentially disastrous invasive species."[4]

There are several good reasons for taking their suggestion seriously. After all, biologists and management officials are in widespread agreement that invasive species are disastrous for local environments. In 1992, the famed Harvard naturalist E. O. Wilson described invasive species as the second biggest threat to biodiversity in the twenty-first century, and scientists have echoed the claim innumerable times for more than two decades.[5] Many biologists now insist that invasive species place more than half the native species in the United States under the threat of extinction. In addition to the ecological reasons, however, there are also economic ones. Some estimate that invasive species cost the U.S. economy more than $100 billion on an annual basis, and that they cost the world economy hundreds of billions more.[6]

And yet, despite this consensus, a growing number of people now question whether invasive species have truly earned their terrible reputation. Echoing similar conclusions from disaster studies, many people now highlight humanity's complicit role in so-called "invasions."[7] Many biologists recognize that human activity drives the invasion process, and some have thus suggested that we relabel "invasive species" as "disturbance specialists."[8] Others are more insistent, denouncing invasion biology as "pseudoscience."[9] They note that the "invasive" label is entirely subjective, and that it is not at all clear what qualifies as "native" versus "invasive." In many (and perhaps all) cases, the "native" population is itself derived from an "invasive" ancestor, and so one can never truly designate any particular time period or ecological condition as a "natural" baseline. Scholars have also shown that invasion biology, the scientific discipline that studies the introduction, spread, and impacts of non-native populations, is sometimes rife with nativism and xenophobia.[10] In his recent book, *American Perceptions of Invasive Species,* historian Peter Coates situates the history of immigrant plants and animals within the wider history of human immigration. He allows that invasion biology might be scientifically sound, but he also writes that the field is not immune from overtly "racist anthropomorphizing." His numerous examples show that Americans have often mobilized racial and ethnic stereotypes in an attempt to disparage non-native populations, human or otherwise.[11]

This essay will assess whether invasive species qualify as disasters, and whether two previously distinct bodies of scholarship (invasion biology and disaster studies) offer anything of service to each other. The essay is divided into three sections. The first provides a brief history of invasion biology that introduces readers to some of the biggest controversies in the field. In particular, we show that there is disagreement among biologists regarding the purportedly "disastrous" nature of biological invasions. The second section focuses on introduced non-native animal populations in the Gulf South. The decision to focus on fauna rather than flora reflects the authors' interests and expertise, but we readily acknowledge that the Gulf South is also home to many different introduced non-native plant species (most notably kudzu) that have helped shape the region's ecology and its identity.[12] The Gulf South is defined as the five states that border the northern Gulf of Mexico: Texas, Louisiana, Mississippi, Alabama, and Florida. There are several reasons for restricting our analysis to this particular region. Many biologists have called the Gulf South a "biodiversity hotspot" because the region contains such a large number of plant and animal species found nowhere else on Earth.[13] Alas, biodiversity hotspots are especially prone to invasions, and, as a result, the Gulf South is home to more invasive species than any other region of the continental United States.[14] The third and final section will compare and contrast the central insights from invasion biology and disaster studies and will assess whether biological invasions qualify as environmental disasters.

A Brief History of Invasion Biology

People often overlook the evolutionary significance of geographical features like deserts, mountains, and oceans, even though these features have influenced the history of life on Earth in profound ways. For millions of years, these physical barriers have served as reproductive barriers, preventing plants and animals from migrating to different parts of the world and thus restricting gene flow. As a result, life incubated within certain well-defined ecological theaters, known as biogeographical realms, and there was relatively little opportunity for the flora and fauna in one arena to migrate into another. That famously changed in the late fifteenth century, when Christopher Columbus and the first generation of transatlantic explorers chanced to "discover" the Western Hemisphere (where millions of people already lived), and thus initiated the greatest biological "reshuffling" in the history of our planet. As explorers and colonizers crisscrossed the globe over the next several centuries, innumerable species were transplanted to new continents, where ecological conditions were often dramatically different. Many of these species were introduced on purpose, though just as many, if not more, were introduced accidentally.[15]

Identifying the earliest scientific accounts of introduced species is not easy, especially since the labels used to describe these populations have changed so much over time. Even so, people have acknowledged the conspicuous ecological impact of non-native species for hundreds of years. As early as the 1620s, English natural philosopher Francis Bacon described numerous different species from the Indies that had been accidentally introduced to European ports. When Swedish botanist Peter Kalm visited Pennsylvania in the 1740s, he identified more than a dozen European plant and animal species that had been introduced to the New World, knowingly or otherwise. Recognizing that the planet's rapidly reshuffling biota warranted new terminologies, English naturalist Hewett Cottrell Watson proposed the labels "native" and "alien" to describe displaced biological entities in 1846. In the decades thereafter, his countryman Alfred Russell Wallace acknowledged the singular success of non-native species when describing the planet's aforementioned biogeographic realms, but he neither celebrated nor condemned their existence.[16]

As appreciation for biodiversity increased in the late nineteenth and early twentieth centuries, scientists adopted increasingly hostile attitudes toward non-native species. The great naturalist Aldo Leopold devoted his life to developing a land ethic that would extend dignity to the rest of nature, but he adopted increasingly negative opinions about non-native species in the 1930s and 1940s. He feared that they would disrupt established energy flows within preexisting biological communities, and that they would decrease biodiversity. Some have suggested that Leopold's position reflected the undercurrents of nativism that were pervasive in the United States at the time. It is true that Leopold believed most "runaway populations" were "foreigners" from distant lands, though he acknowledged native species were equally capable of "pest behavior."[17] Meanwhile, American zoologist Marston Bates described humanity's conspicuous ecological footprint in a 1956 essay titled "Man as an Agent in the Spread of Organisms."[18]

The most influential book in the history of invasion biology is Charles Elton's *Ecology of Invasions by Plants and Animals* (1958). In exceedingly readable prose, Elton describes the three basic components of invasion biology: "the evolution of distinct biotas in isolation, the shattering of that isolation by human trade and travel, and the disastrous impacts of some of this mixing."[19] Significantly, he portrayed non-native species as an existential threat. "It is not just nuclear bombs and wars that threaten us, though these rank very high on the list at the moment: there are other sorts of explosions, and this book is about ecological explosions." On two separate occasions in the book, he explicitly refers to biological invasions as "disasters," and several of his examples are drawn from the Gulf South.[20] Elton's book shaped all debates about non-native species for decades thereafter, and it even helped launch a new field of study known as "invasion biology."[21] American ecologist Daniel Simberloff has gone so far as to call Elton's *Ecology of Invasions* the "Bible of invasion biology."[22] To be sure, most of the subsequent literature has adopted Elton's resolute opinion that non-native species represent a threat and need to be eliminated.[23]

That being said, Elton's legacy is more complicated than many biologists appreciate. Consider, for example, that he recognized humanity's complicit role in biological invasions, and he acknowledged that many "invasive" species had been deliberately imported. What is more, he admitted that most invasions occurred in soils that had been recently disturbed by people. "It will be noticed that invasions most often come to cultivated land, or to land much modified by human practice," he wrote.[24] Contemporary biologists share Elton's belief that humans are invariably complicit in biological invasions. According to Simberloff, non-native species that translocate on their own (via winds, waves, and so forth), without human participation, cannot be "invasive" as such.[25] The invasive label therefore connotes that humans are both the aggrieved party *and* the responsible party. Moreover, many of the biologists who have studied non-native species in Elton's wake have managed to resist his overtly militaristic rhetoric.[26] In fact, the field of invasion biology has inspired a "cottage industry of criticisms" during its relatively brief existence.[27] The field's most extreme critics insist that invasion biology is little more than a "pseudoscience" that is motivated by nativism, xenophobia, and fear.[28] The world's leading invasion biologists have been forced to defend their science and their character. "The goal is not a reduction of numbers of nonindigenous species per se," ecologist David Lodge and philosopher Kristin Shrader-Frechette write, but rather "a reduction in the damage caused by invasive species, including many sorts of environmental and economic damage."[29] Some doubt whether the two sides will ever find common ground. After all, disputes over the legitimacy of invasion biology are really cultural disputes over "differing worldviews."[30]

Introduced Non-Native Animal Populations in the Gulf South

The Gulf South is one of North America's most distinct ecosystems. The five states that comprise the region boast a relatively warm and humid climate that supports diverse plant and animal species. The coastline from southern Texas to southern Florida arcs more than sixteen hundred miles around the northern Gulf of Mexico, but this measurement does not include the coastline's seemingly infinite inlets and waterways, which provide the Gulf South with tens of thousands of extra miles of meandering coastline. Inland from the water's edge, the Gulf Coastal Plain features flat expanses, wide riverine shallows, and numerous marshy wetlands. Habitats range from mangrove swamps to coastal grasslands, from sandy pine barrens to muddy riparian deltas. The broad Mississippi River slices right through the middle of the region and discharges into the Gulf of Mexico. Displaced silt from the continent's vast interior collects at the mouth of the river. These delicate wetlands support diverse forms of life and numerous human industries.[31]

It is appropriate that we should begin our analysis of introduced species in the Gulf South with a discussion on feral pigs (*Sus scrofa*), who are frequently cited as among the worst invasive species in the United States.[32] They forage for food in the soil, generally destroying any landscape they encounter. Their rooting damages forests, wildlife, soil, and water quality, and their parasitic pathogens threaten human lives.[33] Free-ranging pigs have been blamed for the decline and extinction of numerous plant and animal species. Their impact is not merely environmental, however. Officials estimate that the animals cost the nation approximately $1.5 billion in crop damages every year, while others insist that they are responsible for more than $20 billion in damages per year. Feral pigs will live just about anywhere, but they have thrived especially well in the Gulf South. In fact, more than half of the nation's six million free-ranging pigs live in the five Gulf states. Biologists and management officials have waged a "war on pigs" for the past several decades, but the pigs have nevertheless proven almost impossible to eradicate.[34]

Pigs first entered the Gulf South along with Spanish explorer Hernando De Soto in 1539. The thirteen pigs he transported later swelled to several hundred as his army meandered across the Gulf South, from Florida to Texas, for the next several years. Generations of historians have suggested that some of the conquistador's animals might have escaped and established feral populations, but conclusive proof is lacking. In any event, the region's pig population boomed during the early national period, when thousands of people migrated into the Florida peninsula and the Old Southwest alongside their livestock.[35] The open range persisted in the Gulf South long after it had closed in the more densely settled regions of the Northeast and Midwest. As a result, tourist sportsmen and sportswomen who hunted waterfowl in the region were surprised to discover that feral pigs roamed the land and provided thrilling sport.[36]

Pigs did not become invasive, in name or fact, until the twentieth century, when the southern range finally collapsed. Thereafter, pigs were legally prohibited from existing anywhere other than inside a pen. However, closing the commons did not automatically eject the large numbers of feral pigs who inhabited the pine forests, scrub prairies, and riverine shallows. Instead they persisted for years thereafter. When American ecologist Tom McKnight studied the biogeography of feral animals in 1964, he reported around 1.5 million pigs in the United States. He noted that they were concentrated in the Gulf South, and that they were highly destructive.[37] When ecologists Jack Mayer and Lehr Brisbin described feral pigs in 1991, they estimated that approximately two million pigs lived in the United States, and that the vast majority lived in the Gulf South. When Mayer and Brisbin revisited the topic in 2009, they reported that the problem had grown considerably worse. The number of states reporting feral pigs had increased from nineteen to forty-four, while the national population had grown from around 3 million to nearly 6 million. Among this total, well over half (around 4,315,000) lived in the five Gulf states.[38]

In 2014, Congress approved $20 million in funding for the National Feral Swine Damage Management Program, the first time the U.S. Department of Agriculture addressed feral pigs on a national level. Despite the new funds, however, success is by no means guaranteed. Biologists and management officials employ a variety of weapons in their battle against feral pigs, including guns, poisons, and helicopters, but the pig population remains as high as ever. One would think that they could call upon the region's comparatively large number of hunters for help, especially since pigs are now the second most popular hunting targets in the Gulf South, trailing only whitetail deer, but that has not happened.[39] In fact, paradoxical though it may sound, pig hunting is actually making the so-called invasion much worse. Biologists, management officials, and even hunters agree that hunters bear primary responsibility for the population explosion. On numerous occasions, hunters have knowingly translocated pigs so they can have a hunting quarry closer to home. In other words, the desire to hunt pigs is fueling the pig invasion.

Of course, pigs are not the only mammals who have been introduced to the Gulf South. In addition, millions of non-native nutrias live in the region. These large rodents (adults weigh around twenty pounds) are found in all five Gulf states, but they prefer wetland habitats and are therefore most abundant in Louisiana. Officials at the local, state, and national levels regard Louisiana's nutrias as highly invasive.[40] They explain that nutrias overfeed on marsh plants and thus contribute directly toward the destruction of Louisiana's fragile wetlands. When areas with especially high nutria populations are denuded of their vegetation, they eventually convert to open water.[41] Experts estimate that Louisiana has lost about 22,000 acres of marshland to nutrias, and that the animals negatively affect another 100,000 acres. Nutrias are also accused of devastating agriculture by foraging on crops and weakening irrigation structures. They are especially destructive toward Louisiana's most valuable crops—soybeans, rice, and sugarcane—and thus threaten the state's multibillion-dollar agricultural industry.[42]

Although nutrias are "now almost universally reviled," that was not always the case.[43] In fact, prospective pelt farmers in Louisiana deliberately imported the animals in 1937.[44] Popular legend has long held that Edward Avery McIlhenny, head of the Tabasco Company, was responsible for the nutrias' dispersal throughout the wetlands.[45] Shane K. Bernard's research in the company archives has conclusively shown that McIlhenny was not the first person to bring nutrias to Louisiana, and that at least two other farmers acquired nutrias prior to him. In fact, when McIlhenny resolved to raise pelts for the fur industry in 1938, he acquired his first nutrias from farmers in Louisiana. By the following decade, he had discovered that it was far more cost-effective to release the animals into the wild, and so he "liberated" the nutrias into the wetlands in 1945.[46] In the years thereafter, nutrias expanded their range in every direction.[47] Although trappers collected just 18,000 pelts in 1946, they annually collected more than 400,000 pelts by 1956.[48] Even so, this harvest was a small fraction of the actual population, which numbered closer to twenty million in the late 1950s.[49] Between the early 1960s and the early 1980s, trappers annually harvested more than a million nutrias from coastal wetlands.

Global demand for fur sustained these oversized harvests, and thus helped mitigate the animals' impact on Louisiana's wetlands.[50] Things began to change in the late 1980s and 1990s, as fur began falling out of fashion around the world. The dropping demand for pelts led to relaxed anthropogenic selection of Louisiana's nutria populations. Although trappers collected 1.7 million pelts in 1976, they collected fewer than 115,000 pelts a decade later. Populations swelled as harvests decreased.[51] Predictably, reports about nutria-caused damage increased exponentially.[52] By the turn of the century, trappers collected fewer than 30,000 pelts annually, and as a result, the nutria population continued to expand. In 2002, Louisiana implemented the Coastwide Nutria Control Program, which pays hunters and trappers a four-dollar bounty for every animal they collect. By 2006, trappers and hunters were again harvesting approximately 400,000 from the wetlands on an annual basis.[53] In 2015, they collected another 340,000.[54] Meanwhile, untold millions remain at large in the wetlands. Management officials have accepted that nutrias cannot be removed from the region, but they are still eager to limit their numbers.[55]

Mammals are not the only animals who have "invaded" the Gulf South. On the contrary, insects have descended on the region and successfully reshaped its flora and fauna in dramatic fashion. This is especially true of fire ants, who are now widespread throughout the Gulf South and beyond. Many scientists insist that fire ants outcompete and displace native ants, and that they reduce species density on local and biogeographic scales.[56] In addition to the environmental costs, however, there are financial costs. One recent study found that fire ants cost the Texas cattle industry more than $250 million annually.[57] Another study found that fire ants may well cost American citizens more than $6 billion annually in control measures, medical treatment, and property damage.[58] Once again, understanding how these introduced animals "conquered" the Gulf South reveals as much about humans as it does about insects.

Fire ants are native to South America. They first entered North America as stowaways on a ship that reached Mobile, Alabama, in 1918. Another wave of fire ants arrived in Mobile in the 1930s, and it was this later wave that would eventually "invade" the rest of the Gulf South.[59] Throughout the 1940s and 1950s, fire ants expanded their range both east and west. By 1958, fire ants covered more than 62 million acres in the United States, including the entire Gulf South.[60] In an effort to stem the ants' advance, Congress gave the U.S. Department of Agriculture approximately $2.4 million in 1957 to develop an effective eradication campaign using chemicals and pesticides.[61] It soon became clear that the pesticides were not eradicating the fire ants but were killing off just about every other kind of animal in the vicinity. Rachel Carson discussed fire ants at length when she published *Silent Spring* in 1962. Among other things, Carson insisted that fire ants were not at all invasive, writing that "the fire ant has never been a menace to agriculture and that the facts concerning it have been completely misrepresented." She maintained that the reckless overuse of chemical pesticides was the only real disaster associated with the ants. Authorities at the state and federal levels continued using chemical pesticides to attack fire ants in the years thereafter, but they eventually abandoned the practice in the 1970s.[62] The ants had won.

The case of the fire ants provides several lessons. First, Carson's observation that the pesticides were more disastrous than the pests—that the medicine was worse than the illness—was largely proved true. Between the 1940s and 1970s, government officials employed broad-spectrum poisons that killed any and all animals *except* fire ants. As Joshua Blu Buhs recently observed, fire ants simply "reinvaded the poisoned parcels of land, mocking the eradication ideal."[63] Meanwhile, much as humans are ultimately to blame for the disastrous effects of pigs and nutrias, humans are also fueling the fire ants' expansive radiation. After all, fire ants only colonize disturbed habitats, and nothing has disturbed the soils of the Gulf South quite so much as the vast suburbanization that C. Vann Woodward called the "bulldozer revolution."[64] As a result, anthropogenic activities are fueling the fire ants' radiation.[65] As biologists Joshua R. King and Walter R. Tschinkel recently remarked, "Human activity, not biological invasion, is the primary driver of negative effects on native communities and of the process of invasion itself." They hypothesize that fire ants, and, indeed, all invasive species, would be more accurately described as "disturbance specialists."[66] Finally, the Gulf South is not only the scene of fire ant invasions, but also the source. Genetic analyses have shown that most of the world's displaced fire ant populations were not introduced from South America, but rather from the Gulf South. This complicates traditional ideas about invasions, and for that matter, blame.[67]

Other supposedly invasive insects have yielded different lessons. Consider the case of the boll weevil, who began wreaking environmental and economic havoc in the Gulf South more than one hundred years ago. The insect's disastrous impact prompted one federal agent to describe the first boll weevils in the region as a "wave of evil."[68] In 1908, Mississippi Delta planter LeRoy Percy wrote to a friend that "the weevil will bring with him disaster." Despite its devastating impact on the region, some communities hold the boll weevil in high regard. Throughout the Gulf South, people have celebrated boll weevils in artwork, songs, sculptures, and folklore. As Fabian Lange recently remarked, "The boll weevil is America's most celebrated agricultural pest."[69]

To understand this strange state of affairs, one must first understand cotton's unique place in the history of the Gulf South. In the second half of the nineteenth century, the rapid expansion of railroads in the region facilitated the rapid expansion of cotton cultivation. By the early twentieth century, "No region of the country, perhaps the world, was more devoted to mass production of the fleecy white crop." That is what made the earliest reports about boll weevils so alarming. The insects live their entire lives inside cotton bolls, an ecological niche that destroys the plant.[70] Boll weevils first appeared along the southern Texas coast in 1892. Within a few years, total cotton production declined by nearly 50 percent.[71] The insects spread across the Gulf South, advancing between 40 to 160 miles per year.[72] They reached Louisiana by 1903, and Mississippi by 1908.[73] As cotton fields were destroyed, sharecroppers abandoned the fields and moved en masse. As a result, many researchers have credited the boll weevil infestation with triggering a "Great Migration" of African Americans from the Gulf South to urban centers farther north.[74] Some credit boll weevils with vanquishing King Cotton and finally forcing planters to diversify. According to this narrative, boll weevils are "liberators."[75] In 1919, this sentiment inspired the people in Enterprise, Alabama, to erect a statue in the insect's honor.[76]

Yet there are reasons to doubt that narrative. While it is true that some places abandoned cotton cultivation in the wake of the boll weevil onslaught, other places doubled down.[77] In Mississippi, prime ecological conditions allowed planters to withstand the boll weevils' initially devastating impact, and to extract ever more labor from the region's overwhelmingly African American workforce.[78] Indeed, statistics reveal that cotton cultivation increased dramatically in some parts of the Gulf South.[79] Ten years after the people of Enterprise erected a statue honoring the boll weevils' liberating influence, the county's farmers were harvesting just as much cotton as they had before the insects' arrival. In 1929, most farms in Alabama, Mississippi, and Louisiana still received the majority of their profits from cotton.[80] According to this narrative, boll weevils did not challenge the region's devotion to monoculture.

There are several lessons that one can draw from the case of boll weevils. First, the contested history of the boll weevils proves that invasions, including their purportedly disastrous effects, are open to interpretation. Second, farmers in the Gulf South could track the boll weevils' inexorable advance for several years prior to the insects' actual arrival. In this way, the insects represented a hazard for which planters could, in theory, prepare. More frequently, they expanded cotton cultivation, "as if trying to squeeze one last big crop rather than beginning to diversify away from their threatened staple."[81] Finally, the current status of boll weevils is also instructive. Since the 1980s, the USDA has administered a campaign that has almost completely eradicated boll weevils from the Gulf South (scattered pockets remain in southern Texas). There is a sense that the invasion has been thwarted—that the disaster has been successfully navigated—but those conclusions are premature.

Animals known as exotics have also moved into the region recently, and none have generated as much publicity as the Burmese python. "The python invasion may rival all others in terms of its potential to completely alter the structure of native ecosystems and to capture the public's attention," biologists Michael E. Dorcas and J. D. Willson write. They add: "Invasive pythons in the United States have become an environmental specter that urgently warrants public concern." How did all of these pythons come to populate the region? As their name suggests, Burmese pythons are native to Southeast Asia. They might have remained there, but the growing demand for exotic pets ensured that some were plucked from their homeland and shipped to traders in the United States. In the late twentieth century, business was booming. Between 1989 and 2000, traders imported more than 400,000 snakes, including more than 100,000 Burmese pythons. The vast majority of these live shipments passed through South Florida, which was the nexus of the international exotic-pet trade.[82]

To be sure, these facts may explain how pythons made it to South Florida, but they do not explain how they escaped into the Everglades. There are many theories about how the snakes got there, but the truth is that researchers do not know if the first snakes were intentionally released or whether they somehow escaped from captivity. The earliest reports of pythons in Everglades National Park date back to the late

1990s, though sightings increased dramatically in the years thereafter. In 2004, biologist Skip Snow reported that four pythons recently taken from the Everglades contained thirty-five eggs among them.[83] The rising number of pythons began receiving international attention in 2005, when news outlets around the world circulated a photograph taken in the Everglades. The photograph showed a dead python with a dead alligator protruding from the snake's stomach. For many, the grim scene neatly encapsulated the tension between a great many native and nonnative species.[84] Later that same year, biologist Robert N. Reed warned that pythons could easily establish populations in the Florida wilds.[85] Sure enough, pythons began appearing in the park with even greater frequency. In 2006, scientists discovered several python nests within the boundaries of the national park.

As of late 2015, more than two thousand pythons have been removed from the park and surrounding areas. Officials are generally cautious when calculating the total number of pythons living in the Everglades, but estimates range from ten thousand to thirty thousand. The pythons are not going anywhere, and, as a result, the park's fauna looks considerably different from how it appeared just a few short years ago. Several scientific studies have confirmed that the rising number of pythons has led to fewer mammals and birds.[86] Meanwhile, the pythons' explosive population growth has prompted a spirited debate in the scientific literature over their potential range expansion.[87] Understandably, the debate is growing more urgent. Pet pythons have killed several children in Florida over the past twenty years.[88] Wild pythons have not yet claimed any human lives, but that has hardly quelled fears.[89] State and federal wildlife officials have spent millions of dollars trying to combat the growing python infestation, and they have utilized a variety of tools and methods, including traps, bounties, and organized hunts, yet these efforts have so far failed to produce substantive results.[90] Further complicating matters, state and federal agencies have spent more than $100 million rebuilding stork and muskrat populations, yet Burmese pythons feed on both of these species voraciously.[91] Despite everything, even the pythons have their defenders. Dorcas and Willson have argued, "In general, snakes have an undeserved bad reputation. Like most other animals, snakes are generally afraid of people."[92] Indeed, it is hard to blame the pythons for their invasion when humans quite clearly were responsible.

Other exotic animals have invaded not only the lands bordering the Gulf of Mexico, but the Gulf itself. Perhaps no aquatic invader is more notorious than the lionfish. Native to Indo-Pacific waters, these visually striking fish were first transported to the Western Hemisphere to serve as ornamental pets. Some were released from their aquaria and established populations in the wild. Lionfish first arrived in North American waters in 1985, when fishermen spotted them on the east coast of Florida.[93] They expanded their range up the Atlantic Coast in the years thereafter, and were first seen in the Florida Keys in 2009.[94] By 2010, they were being identified in the northern Gulf of Mexico, and they are now found off the coast of all five Gulf states.[95] Meanwhile, recent research has shown that lionfish are not only taking over reefs, but also growing more abundant at greater depths.[96]

There is widespread consensus that lionfish have a negative ecological impact, not least because they prey on already critically endangered reef fish.[97] Accordingly, in addition to the ecological implications, there is an economic impact as lionfish disrupt existing ecosystems and food webs, and thus threaten the seafood industry of the Gulf South. Struggling to respond to this invasion as it happens, officials have suggested a variety of control measures. Some think we should import still more voracious species to feed on the lionfish.[98] Others think that *people* ought to start eating them, and many are now promoting the fish as edible fare. These inventive efforts notwithstanding, many scientists and officials have resigned themselves to defeat.[99]

Recent events suggest that the threat from exotic populations will only grow more acute in the future. For example, researchers in Florida, which has the largest number of introduced amphibians and reptiles in the world, have recently confirmed that the Nile crocodile (*Crocodylus niloticus*) is now living in the state, though we do not know for certain how many there are.[100] Over the last ten years, many Nile crocodiles have been imported from South Africa and Madagascar for display in zoos, or as part of the exotic-pet trade. Much like the Burmese python, it is the latter that most likely explains the introduction of the Nile crocodile to Florida.[101] If these animals become established in their new home, they will pose a serious threat both to local fauna and to human beings. Unlike their gentler cousin, the American crocodile, who grows to four meters and is satisfied with fish, crustaceans, and turtles, the Nile crocodile grows to six meters, weighs over one and a half tons (it is the second largest reptile on earth), and is known to eat large mammals such as zebras, hippos, and humans. In 2015 alone, there were eighty-eight documented attacks against human beings in Africa, fifty-eight of which were fatal.[102] While there is no cause for fear or panic at this point, we have yet to see what the effects will be, if any, of this more recently introduced species in the Gulf South.

Conclusion: Biological Invasions as Disasters

Disasters are generally measured in three different ways: economic costs, environmental devastation, and loss of human life. We must admit that this lattermost metric—loss of human life—scarcely factors into debates over biological invasions. Earthquakes and plagues routinely kill hundreds, sometimes thousands of people, but even the most pernicious invasive species do not threaten human lives. For all their terrifying ubiquity, the Burmese pythons inhabiting the Everglades have not yet claimed any human victims. Thus, when locals remark that the Gulf South has been "biologically traumatized" by non-native species, one must take the observation with a grain of salt.[103] In this way, we must allow that biological invasions are not like disasters, and that the stakes are always much higher when human lives are at risk.

That important qualification out of the way, it is equally important to highlight the numerous ways in which biological invasions and disasters *do* resemble one another. Both categories of events pose unique threats to the ecosystems and, indeed, the very landscapes of the Gulf South. Like so many abiotic disasters, non-native species destroy enormous swaths of fragile wetlands, and as Roberto Barrios and Kevin Fox Gotham both explain in their respective contributions to this volume, the destruction of wetlands most seriously affects those people who derive their livelihood and identity from the wetlands. In other words, biological invasions and natural hazards both disproportionately affect the most vulnerable. Meanwhile, the ways in which researchers conceptualize these events have also changed. For instance, disasters were once regarded as discrete events. As Andy Horowitz explains elsewhere in this volume, "many writers continue to narrate so-called natural disasters as acute events that erupt in a catastrophic instant." In similar fashion, biologists have long regarded invasions as comparatively sudden phenomena. Elton even likened them to explosions. "I use the word 'explosion' deliberately," he explained, "because it means the bursting out from control of forces that were previously held in restraint by other forces."[104] In both fields, however, practitioners have begun to adopt more nuanced views. Researchers now agree that the worst effects of many so-called biological invasions unfold over years, decades, and in some cases, centuries—an example, in other words, of a slow disaster.

Another conspicuous similarity between biological invasions and disasters concerns humanity's invariably complicit role in both. Just as humans are frequently to blame for the worst aspects of disasters, so humans are likewise to blame for the most disastrous effects of nonnative species. Put differently, just as each disaster lays bare existing social structures and vulnerabilities, each biological invasion reveals something different about humanity's complicit role in disasters. The desire to hunt and to kill paradoxically fuels the region's booming population of feral pigs. Changing tastes in fashion ultimately explain the explosive growth of nutrias in the Gulf South. The rapid expansion of fire ants would have never been possible without people first sprawling into previously undisturbed habitats. In similar fashion, neither pythons nor lionfish would inhabit the Gulf South if not for humanity's passion for collecting dangerous novelties and exotic beauties, respectively. Meanwhile, boll weevils have been largely eradicated from the region, but they remain a threat to communities dependent on cotton production.

Given humanity's complicit role in all biological invasions, it should come as no surprise that some regard humans as the most invasive species of all. The famed American conservationist George Perkins Marsh was among the first to focus on humanity's disruptive tendencies. "Man is everywhere a disturbing agent," he wrote in the early twentieth century, adding, "wherever he plants his foot, the harmonies of nature are turned to discords."[105] Contemporary researchers are more explicit still. Some cite humanity's exponential growth and unsustainable behavior as evidence that our species is destroying the rest of the planet and that we collectively behave like a malignant tumor.[106] Others note that humans are driving one of the largest extinction events in the long history of the planet.[107] According to this view, invasive species are not the cause of disaster but rather the instrument, and biological translocations will continue to increase as globalization accelerates. Many fear that biomes and cultures will grow increasingly "homogenized" in the years ahead.[108] Elton and Alfred Crosby both warned about this possibility, and others have since followed suit. "The earth is hurtling towards one world culture and (maybe) one world ecosystem," Australian biologist Timothy Low writes.[109] Meanwhile, scientists predict that anthropogenic climate change will facilitate even more species displacement in the years ahead.[110] But much can be done to help slow this disruptive process.

Like the other essayists in this volume, who suggest ways to increase resilience following disasters, we too would like to offer a few suggestions. While some might prefer to prevent immigration altogether, we find that position unwise and untenable. Instead, we assert that increasing resilience requires that *we* change. It requires that humans accept responsibility for the biotic composition of the planet, and that we proceed with deliberate care when cultivating and pruning our shared future. Since we are looking for guidance, it is perhaps fitting that we close by revisiting the "Bible of invasion biology," Charles Elton's *Ecology of Invasions*. Even though Elton is now most frequently remembered for his militaristic attitude toward non-native species, he offers surprisingly measured advice in the book's final chapter: "From now on, it is vital that everyone who feels inclined to change or cut away or drain or spray or plant any strip or corner of the land should ask themselves three questions: what animals and plants live in it, what beauty and interest may be lost, and what extra risk changing it will add to the accumulating instability of communities."[111] Inhabitants of the Gulf South may not have it within their power to remove or control non-native species, but that does not mean that they cannot forestall disasters.

NOTES

1. Ted Steinberg, *Acts of God: The Unnatural History of Natural Disaster in America* (New York: Oxford University Press, 2000); Robert A. Stallings, "Causality and 'Natural' Disasters," *Contemporary Sociology* 35 (May 2006): 223–27; Ernest Zebrowski Jr., *Perils of a Restless Planet: Scientific Perspectives on Natural Disaster* (Cambridge, U.K.: Cambridge University Press, 1997).

2. An enormous number of scholarly works examine abiotic disasters. Examples include: Jelle Zeilinga de Boer et al., *Earthquakes in Human History: The Far-Reaching Effects of Seismic Effects of Seismic Disruptions* (Princeton, N.J.: Princeton University Press, 2005); Alvaro S. Pereira, "The Opportunity of a Disaster: The Economic Impact of the 1755 Lisbon Earthquake," *Journal of Economic History* 69 (June 2009): 466–99; Charles F. Walker, *Shaky Colonialism: The 1746 Earthquake-Tsunami in Lima, Peru, and Its Long Aftermath* (Durham, N.C.: Duke University Press, 2008); Gregory Clancy, *Earthquake Nation: The Cultural Politics of Japanese Seismicity, 1868–1930* (Berkeley: University of California Press, 2006); Susan Millar Williams and Stephen G. Hoffius, *Upheaval in Charleston: Earthquake and Murder on the Eve of Jim Crow* (Athens: University of Georgia Press, 2011); Deborah R. Coen, *The Earthquake Observers: Disaster Science from Lisbon to Richter* (Chicago: University of Chicago Press, 2012); Matthew Mulcahy, *Hurricanes and Society in the British Greater Caribbean, 1624–1783* (Baltimore: Johns Hopkins University Press, 2006); Geoffrey Parker, *Global Crisis: War, Climate Change and Catastrophe in the Seventeenth Century* (New Haven, Conn.: Yale University Press, 2013); Steinberg, *Acts of God*; Scott Gabriel Knowles, *The Disaster Experts: Mastering Risk in Modern America* (Philadelphia: University of Pennsylvania Press, 2011); Naomi Klein, *The Shock Doctrine: The Rise of Disaster Capitalism* (New York: Knopf, 2007); Christof Mauch and Christian Pfister, eds., *Natural Disasters, Cultural Responses: Case Studies toward a Global Environmental History* (Lanham, Md.: Lexington Books, 2009). Numerous works examine diseases within the context of disaster studies, including, but not limited to, Henry M. McKiven Jr., "The Political Construction of a Natural Disaster: The Yellow Fever Epidemics of 1853," *Journal of American History* 94 (December 2007): 734–42; Dorothee Brantz, "'Risky Business': Disease, Disaster and the Unintended Consequences of Epizootics in Eighteenth- and Nineteenth-Century France and Germany," *Environment and History* 17 (2011): 35–51; William H. McNeill, *Plagues and Peoples* (New York: Anchor Books, 1976).

3. Angus Macleod Gunn, *Encyclopedia of Disasters: Environmental Catastrophes and Human Tragedies* (Westport, Conn.: Greenwood Press, 2008).

4. Anthony Ricciardi et al., "Should Biological Invasions Be Managed as Natural Disasters?" *BioScience* 61 (April 2011): 312–17.

5. Matthew K. Chew, "Ecologists, Environmentalists, Experts, and the Invasion of the 'Second Greatest Threat,'" *International Review of Environmental History* 1 (2015): 7–40.

6. David Pimental et al., "Update on the Environmental and Economic Costs Associated with Alien-Invasive Species in the United States," *Ecological Economics* 52 (2005): 273–88; Daniel Simberloff, *Invasive Species: What Everyone Needs to Know* (Oxford, U.K.: Oxford University Press, 2013); Daniel Simberloff, "Integrity, Stability, and Beauty: Aldo Leopold's Evolving View of Nonnative Species," *Environmental History* 17 (July 2012): 487–511; Mick N. Clout and Peter A. Williams, *Invasive Species Management: A Handbook of Principles and Techniques* (Oxford, U.K.: Oxford University Press, 2009); Daniel Simberloff and Marcel Rejmanek, *Encyclopedia of Biological Invasions* (Berkeley: University of California Press, 2011); Susan L. Woodward and Joyce A. Quinn, *Encyclopedia of Invasive Species: From Africanized Honey Bees to Zebra Mussels* (Santa Barbara, Calif.: ABC-CLIO, 2011); David Pimental, ed., *Biological Invasions: Economic and Environmental Costs of Alien Plant, Animal, and Microbe Species* (Boca Raton, Fla.: CRC Press, 2011); Mark A. Davis, *Invasion Biology* (Oxford, U.K.: Oxford University Press, 2009); Alan Burdick, *Out of Eden: An Odyssey of Ecological Invasion* (New York: Farrar, Straus and Giroux, 2005); J. A. McNeely, ed., *The Great Reshuffling: Human Dimensions of Invasive Alien Species* (Cambridge, U.K.: IUCN, 2001); Elizabeth A. Choresky and John Randall, "The Threat of Invasive Alien Species to Biological Diversity: Setting a Future Course," *Annals of the Missouri Botanical Garden* 90 (Winter 2003): 67–76.

7. Reuben P. Keller et al., eds., *Bioeconomics of Invasive Species: Integrating Ecology, Economics, Policy, and Management* (Oxford, U.K.: Oxford University Press, 2009); Charles Perrings et al., *Bioinvasions and Globalization: Ecology, Economics, Management, and Policy* (Oxford, U.K.: Oxford University Press, 2010).

8. Joshua R. King and Walter R. Tschinkel, "Experimental Evidence That Human Impacts Drive Fire Ant Invasions and Ecological Change," *Proceedings of the National Academy of Sciences* 105 (December 23, 2008): 20339.

9. David I. Theodoropoulos, *Invasion Biology: Critique of a Pseudoscience* (Blythe, Calif.: Avvar Books, 2003).

10. Daniel Simberloff, "Confronting Introduced Species: A Form of Xenophobia?" *Biological Invasions* 5 (2003): 179–92; Daniel Simberloff and Jean R. S. Vitule, "A Call for an End to Calls for the End of Invasion Biology," *Oikos*, 2013, 1–5; Marcel Rejmanek et al., "Biological Invasions: Politics and the Discontinuity of Ecological Terminology," *Bulletin of the Ecological Society of America* 83 (April 2002); Ian D. Rotherham and Robert A. Lambert, eds., *Invasive and Introduced Plants and Animals: Human Perceptions, Attitudes and Approaches* (Washington, D.C.: Earthscan, 2011); Seth R. Reice, *The Silver Lining: The Benefits of Natural Disasters* (Princeton, N.J.: Princeton University Press, 2001); Erika S. Zavaleta et al., "Viewing Invasive Species Removal in a Whole-Ecosystem Context," *Trends in Ecology & Evolution* 16 (August 2001): 454–59; Emma Marris, *Rambunctious Garden: Saving Nature in a Post-Wild World* (New York: Bloomsbury, 2013);

Theodoropoulos, *Invasion Biology;* C. R. Veitch and M. N. Clout, eds., *Turning the Tide: The Eradication of Invasive Species* (Cambridge, U.K.: IUCN, 2002).

11. Peter Coates, *American Perceptions of Immigrant and Invasive Species* (Berkeley: University of California Press, 2007), 4.

12. Richard J. Blaustein, "Kudzu's Invasion into Southern United States Life and Culture," in *The Great Reshuffling,* ed. McNeely, 55–62; Mart Allen Stewart, "Georgia History in Pictures," *Georgia Historical Quarterly* 81 (Spring 1997): 151–67.

13. Eugene P. Odum, "The Southeastern Region: A Biodiversity Haven for Naturalists and Ecologists," *Southeastern Naturalist* 1 (2002): 1–2; Charles Lydeard and Richard L. Mayden, "A Diverse and Endangered Aquatic Ecosystem of the Southeast United States," *Conservation Biology* 9 (August 1995): 800–805; David M. Olsen and Eric Dinerstein, "The Global 200: Priority Ecoregions for Global Conservation," *Annals of the Missouri Botanic Garden* 89 (Spring 2002): 199–224; J. Baird Callicott et al., "Biocomplexity and Conservation of Biodiversity Hotspots: Three Case Studies from the Americas," *Philosophical Transactions: Biological Sciences* 362 (February 28, 2007): 321–33; Sean P. Graham et al., "An Overlooked Hotspot? Rapid Biodiversity Assessment Reveals a Region of Exceptional Herpetofaunal Richness in the Southeastern United States," *Southeastern Naturalist* 9 (2010): 19–34; Richard J. Blaustein, "Biodiversity Hotspot: The Florida Panhandle," *BioScience* 58 (October 2008): 784–90; Paul A. Keddy et al., "Wet and Wonderful: The World's Largest Wetlands Are Conservation Priorities," *BioScience* 59 (January 2009): 39–51.

14. J. D. Fridley et al., "The Invasion Paradox: Reconciling Pattern and Process in Species Invasions," *Ecology* 88 (January 2007): 3–17; John M. Barry, *Rising Tide: The Great Mississippi Flood of 1927 and How It Changed America* (New York: Simon and Schuster, 1997); Erik Larson, *Isaac's Storm: A Man, a Time, and the Deadliest Hurricane in History* (New York: Crown, 1999); Patty Glick et al., "Potential Effects of Sea-Level Rise on Coastal Wetlands in Southeastern Louisiana," *Journal of Coastal Research,* Spring 2013, 211–33; Michael S. Ross et al., "Disturbance and the Rising Tide: The Challenge of Biodiversity Management on Low-Island Ecosystems," *Frontiers in Ecology and the Environment* 7 (November 2009): 471–78.

15. Alfred Crosby, *The Columbian Exchange: Biological and Cultural Consequences of 1492* (Westport, Conn.: Greenwood, 1972); Alfred Crosby, *Ecological Imperialism: The Biological Expansion of Europe, 900–1900* (Cambridge, U.K.: Cambridge University Press, 1986); Jared Diamond, *Guns, Germs, and Steel: The Fates of Human Societies* (New York: Norton, 1997). Regarding biogeographical realms, see David Quammen, *The Song of the Dodo: Island Biogeography in an Age of Extinctions* (New York: Scribner, 1996).

16. Matthew Chew, "Invasion Biology: Historical Precedents," in *Encyclopedia of Biological Invasions,* ed. Simberloff and Rejmanek, 370–72.

17. Simberloff, "Integrity, Stability, and Beauty," 487–511; Matthew K. Chew, "Ending with Elton: Preludes to Invasion Biology" (Ph.D. diss., Arizona State University, 2006).

18. Martson Bates, "Man as an Agent in the Spread of Organisms," in *Man's Role in Changing the Face of the Earth,* ed. William L. Thomas (Chicago: University of Chicago Press, 1956).

19. Daniel Simberloff, "A Rising Tide of Species and Literature: A Review of Some Recent Books on Biological Invasions," *BioScience* 54 (March 2004): 249.

20. Charles Elton, *Ecology of Invasions by Plants and Animals* (1958; Chicago: University of Chicago Press, 2000), 15, 20, 27.

21. Sarah Hayden Reichard and Peter S. White, "Invasion Biology: An Emerging Field of Study," *Annals of the Missouri Botanical Garden* 90 (Winter 2003): 64–66.

22. Simberloff, "A Rising Tide of Species and Literature," 249.

23. Yvonne Baskin, *A Plague of Rats and Rubbervines: The Growing Threat of Species Invasions* (Washington, D.C.: Island Press, 2002); Tim Low, *Feral Future: The Untold Story of Australia's Exotic Invaders* (Chicago: University of Chicago Press, 1999); Jason Van Driesche and Roy Van Driesche, *Nature Out of Place: Biological Invasions in the Global Age* (Washington, D.C.: Island Press, 2000).

24. Elton, *Ecology of Invasions,* 51, 63.

25. Daniel Simberloff, "Biological Invasions: Impacts, Management, and Controversies," in *Controversies in Science and Technology: From Sustainability to Surveillance,* ed. Daniel Lee Kleinman, Karen A. Cloud-Hansen, and Jo Handelsman (Oxford, U.K.: Oxford University Press, 2014), vol. 4: 211–27.

26. Herbert G. Baker and G. Ledyard Stebbins, eds., *The Genetics of Colonizing Species* (New York: Academic Press, 1965).

27. David M. Richardson and Anthony Ricciardi, "Misleading Criticisms of Invasion Science: A Field Guide," *Diversity and Distributions* 19 (2013): 1461–67.

28. Theodoropoulos, *Invasion Biology;* Jonah Peretti, "Nativism and Nature: Rethinking Biological Invasion," *Environmental Values* 7 (May 1998): 183–92; Michael Pollan, "Against Nativism," *New York Times Magazine,* May 15, 1994, 52–55; Banu Subramaniam, "The Aliens Have Landed! Reflections on the Rhetoric of Biological Invasions," *Meridians* 2 (2001): 26–40.

29. David M. Lodge and Kristin Shrader-Frechette, "Nonindigenous Species: Ecological Explanation, Environmental Ethics, and Public Policy," *Conservation Biology* 17 (February 2003): 31.

30. Daniel Simberloff, "Nature, Natives, Nativism, and Management," *Environmental Ethics* 31 (Spring 2012): 5–36; Mark A. Davis, "Invasion Biology, 1958–2004: The Pursuit of Science and Conservation," in *Conceptual Ecology and Invasion Biology: Reciprocal Approaches to Nature,* ed. M. W. Cadotte et al. (London: Kluwer, 2005), xx; Mark A. Davis and Ken Thompson, "Eight Ways to Be a Colonizer, Two Ways to Be an Invader: A Proposed Nomenclature Scheme for Invasion Ecology," *Bulletin of the Ecological Society of America* 81 (July 2000): 226–30; Robert I. Colautti and Hugh J. MacIsaac, "A Neutral Terminology to Define 'Invasive' Species," *Diversity and Distributions* 10 (March 2004): 135–41; Brendon M. H. Larson, "The War of the Roses: Demilitarizing Invasion Biology," *Frontiers in Ecology and the Environment* 3 (2005): 495–500.

31. William J. Mitsch and James G. Gosselink, *Wetlands* (5th ed.; New York: Wiley, 2015); Ann Vileisis, *Discovering the Unknown Landscape: A History of America's Wetlands* (Washington, D.C.: Island Press, 1999).

32. Darryl Fears, "The Dirty Dozen: 12 of the Most Destructive Invasive Animals in the United States," *Washington Post,* February 23, 2015. Readers will notice that we refer to "animals who," rather than "animals that." This distinction reflects the fact that they are not lifeless automata, and are instead sentient organisms whose actions have influenced the course of life on earth. See Abraham H. Gibson, *Feral Animals in the American South: An Evolutionary History* (New York: Cambridge University Press, 2016).

33. Meredith L. McClure et al., "Modeling and Mapping the Probability of Occurrence of Invasive Wild Pigs across the Contiguous United States," *PLoS ONE,* August 12, 2015, 1.

34. Will Brantley, "The Pig Report," *Field & Stream,* September 2015, 78–82.

35. James David Miller, *South by Southwest: Planter Emigration and Identity in the Slave South* (Charlottesville: University of Virginia Press, 2002), 104; Thomas D. Clark and John D. W. Guice, *The Old Southwest, 1795–1830: Frontiers in Conflict* (Albuquerque: University of New Mexico Press, 1989), 112; John Hebron Moore, *The Emergence of the Cotton Kingdom in the Old Southwest: Mississippi, 1770–1860* (Baton Rouge: Louisiana State University Press, 1988), 23.

36. Leonidas Hubbard Jr., "Hunting Wild Hogs as a Sport," *New York Times,* September 1, 1901, sm9; Charles Sloan Reid, "Swine Department: The Mast Hog," *National Stockman and Farmer,* December 5, 1895, 4; "Hunting Wild Hogs," *Dona Ana County Republican* (Las Cruces, N.M.), December 21, 1901, 2.

37. Tom McKnight, *Feral Animals in Anglo-America* (Berkeley: University of California Press, 1964), 43.

38. John J. Mayer and I. Lehr Brisbin Jr., *Wild Pigs: Biology, Damage, Control Techniques* (Aiken, S.C.: Savannah River National Laboratory, 2009), 1.

39. Brantley, "The Pig Report," 81–82.

40. Karen Klima and Steven E. Travis, "Genetic Population Structure of Invasive Nutria (*Myocastor coypus*) in Louisiana, USA: Is It Sufficient for the Development of Eradication Units?" *Biological Invasions* 14 (2012): 1909; Susan Jojola et al., "Nutria: An Invasive Rodent Pest or Valued Resource?" *Wildlife Damage Management Conferences Proceedings* 110 (2005): 120–26; Shane K. Bernard, "M'sieu Ned's Rat? Reconsidering the Origin of Nutria in Louisiana," *Louisiana History* 43 (Summer 2002): 281–93.

41. J. Marx et al., *Nutria Harvest Distribution, 2003–2004, and a Survey of Nutria Herbivory Damage in Coastal Louisiana in 2004* (New Iberia, La.: Fur and Refuge Division, Louisiana Department of Wildlife and Fisheries, 2004); John Baroch et al., "Nutria (*Myocastor coypus*) in Louisiana," *Other Publications in Wildlife Management* 46 (2002): 69–70.

42. Jojola et al., "Nutria," 120–26.

43. Bernard, "M'sieu Ned's Rat?" 282.

44. Frank G. Ashbrook, "Nutrias Grow in United States," *Journal of Wildlife Management* 12 (January 1948): 90.

45. Noel Kinler, Greg Linscombe, and Steve Hartley, *A Survey of Nutria Herbivory Damage in Coastal Louisiana in 1998* (Baton Rouge: Louisiana Department of Wildlife and Fisheries, Fur and Refuge Division [Nutria Harvest and Wetland Demonstration Project], December 31, 1998), 2.

46. Bernard, "M'sieu Ned's Rat?" 281–93; Klima and Travis, "Genetic Population Structure of Invasive Nutria," 1909–18.

47. Ashbrook, "Nutrias Grow in United States," 90.

48. Bernard, "M'sieu Ned's Rat?" 291; Kinler et al., *A Survey of Nutria Herbivory Damage in Coastal Louisiana in 1998*, 2.

49. Baroch et al., "Nutria (*Myocastor coypus*) in Louisiana," 25.

50. "Rodent Nutria Becoming Pest in Southern US," *Science News-Letter* 83 (January 12, 1963): 29; Kinler et al., *A Survey of Nutria Herbivory Damage in Coastal Louisiana in 1998*, 2–3.

51. Baroch et al., "Nutria (*Myocastor coypus*) in Louisiana," 28, 92; Kinler et al., *A Survey of Nutria Herbivory Damage in Coastal Louisiana in 1998*, 3.

52. Kinler et al., *A Survey of Nutria Herbivory Damage in Coastal Louisiana in 1998*, 3.

53. Cheikhna Dedah et al., "The Role of Bounties and Human Behavior on Louisiana Nutria Harvests," *Journal of Agricultural and Applied Economics* 42 (February 2010): 134; Susan M. Jojola et al., "Evaluation of Attractants to Improve Trapping Success of Nutria on Louisiana Coastal Marsh," *Journal of Wildlife Management* 73 (November 2009): 1414.

54. Coastal and Nongame Resources, Louisiana Wildlife & Fisheries, *Coastwide Nutria Control Program, 2014–2015*, July 30, 2015, www.nutria.com.

55. Klima and Travis, "Genetic Population Structure of Invasive Nutria," 1916–18; Susan Pasko and Jason Goldberg, "Review of Harvest Incentives to Control Invasive Species," *Management of Biological Invasions* 5 (2014): 266.

56. Nicholas J. Gotelli and Amy E. Arnett, "Biogeographic Effects of Red Fire Ant Invasion," *Ecology Letters* 3 (2000): 257–61.

57. Steve Teal et al., "The Cost of Red Imported Fire Ant Infestation: The Case of the Texas Cattle Industry," *Texas Journal of Agriculture and Natural Resources* 12 (1999): 88.

58. Marina S. Ascunce et al., "Global Invasion History of the Fire Ant *Solenopsis invicta*," *Science* 331 (February 25, 2011): 1066.

59. Walter Tschinkel, *The Fire Ants* (Cambridge, Mass.: Harvard University Press, 2006), 25; Joshua Blu Buhs, "The Fire Ant Wars: Nature and Science in the Pesticide Controversies of the Late Twentieth Century," *Isis* 93 (2002): 380–81; Kenneth G. Ross and D. DeWayne Shoemaker, "Estimation of the Number of Founders of an Invasive Pest Insect Population: The Fire Ant *Solenopsis invicta* in the USA," *Proceedings: Biological Sciences* 275 (October 7, 2008): 2231.

60. Anne-Marie A. Calcott and Homer L. Collins, "Invasion and Range Expansion of Imported Fire Ants in North America from 1918–1995," *Florida Entomologist* 79 (June 1996): 240–51.

61. Tschinkel, *The Fire Ants*, 45; Buhs, "The Fire Ant Wars," 387.

62. Buhs, "The Fire Ant Wars," 395.

63. Ibid., 391–92, 395–96.

64. C. Vann Woodward, *The Burden of Southern History* (Baton Rouge: Louisiana State University Press, 1993), 6; Matthew D. Lassiter and Kevin M. Kruse, "The Bulldozer

Revolution: Suburbs and Southern History since World War II," *Journal of Southern History* 75 (August 2009): 691–706.

65. B. W. Taylor and R. E. Irwin, "Linking Economic Activities to the Distribution of Exotic Plants," *Proceedings of the National Academy of Sciences* 101 (2004): 17725–30; J. R. King et al., "A Case Study of Human Exacerbation of the Invasive Species Problems: Transport and Establishment of Polygyne Fire Ants in Tallahassee, Florida," *Biological Invasions* 10 (2008).

66. King and Tschinkel, "Experimental Evidence That Human Impacts Drive Fire Ant Invasions and Ecological Change," 20339.

67. Ascunce et al., "Global Invasion History of the Fire Ant *Solenopsis invicta*," 1066; Buhs, "The Fire Ant Wars," 381–82.

68. Fabian Lange et al., "The Impact of the Boll Weevil, 1892–1932," *Journal of Economic History* 69 (September 2009): 685, 718.

69. Ibid., 685.

70. James C. Giesen, "'The Truth about the Boll Weevil': The Nature of Planter Power in the Mississippi Delta," *Environmental History* 14 (October 2009): 683, 684, 688.

71. Lange et al., "The Impact of the Boll Weevil," 708; Kent Osband, "The Boll Weevil versus 'King Cotton,'" *Journal of Economic History* 45 (September 1985): 627.

72. Lange et al., "The Impact of the Boll Weevil," 688.

73. Kathryn Holland Braund, "'Hog Wild,' and 'Nuts': Billy Boll Weevil Comes to the Alabama Wiregrass," *Agricultural History* 63 (Summer 1989): 16; Giesen, "'The Truth about the Boll Weevil,'" 689.

74. Lange et al., "The Impact of the Boll Weevil," 686; Arvarh E. Strickland, "The Strange Affair of the Boll Weevil: The Pest as Liberator," *Agricultural History* 68 (Spring 1994): 161.

75. Braund, "'Hog Wild,' and 'Nuts,'" 15–39; Strickland, "The Strange Affair of the Boll Weevil," 157–68.

76. Osband, "The Boll Weevil versus 'King Cotton,'" 627–43; Strickland, "The Strange Affair of the Boll Weevil," 157–68.

77. Lange et al., "The Impact of the Boll Weevil," 709.

78. Giesen, "'The Truth about the Boll Weevil,'" 683.

79. Pasko and Goldberg, "Review of Harvest Incentives to Control Invasive Species," 263–77.

80. Osband, "The Boll Weevil versus 'King Cotton,'" 628.

81. Lange et al., "The Impact of the Boll Weevil," 715.

82. Michael E. Dorcas and John D. Willson, *Invasive Pythons in the United States: Ecology of an Introduced Predator* (Athens: University of Georgia Press, 2011), 1–3.

83. Kenneth L. Krysko, "Reproduction of the Burmese Python in Southern Florida," *Applied Herpetology* 5 (2008): 93–95.

84. Dorcas and Willson, *Invasive Pythons in the United States,* 2.

85. Robert N. Reed, "An Ecological Risk Assessment of Nonnative Boas and Pythons as Potentially Invasive Species in the United States," *Risk Analysis* 25 (2005): 753–66.

86. Dorcas and Willson, *Invasive Pythons in the United States*, 6; Carla J. Dove et al., "Birds Consumed by the Invasive Burmese Python in Everglades National Park, Florida, USA," *Wilson Journal of Ornithology* 123 (March 2011): 126–31; Michael E. Dorcas et al., "Severe Mammal Declines Coincide with Proliferation of Invasive Burmese Pythons in Everglades National Park," *Proceedings of the National Academy of Sciences* 109 (February 14, 2012): 2418–22; Robert A. McCleery et al., "Marsh Rabbit Mortalities Tie Pythons to the Precipitous Decline of Mammals in the Everglades," *Proceedings of the Royal Society* 282 (2015): 1–7.

87. R. Alexander Pyron et al., "Claims of Potential Expansion throughout the U.S. by Invasive Python Species Are Contradicted by Ecological Niche Models," *PLoS ONE* 3 (August 2008): 1–7; Gordon H. Rodda et al., "What Parts of the US Mainland are Climatically Suitable for Invasive Alien Pythons Spreading from Everglades National Park?" *Biological Invasions* 11 (2009): 241–52; Michael L. Avery et al., "Cold Weather and the Potential Range of Invasive Burmese Pythons," *Biological Invasions* 12 (2010): 3649–52; Michael E. Dorcas et al., "Can Invasive Burmese Pythons Inhabit Temperate Regions of the Southeastern United States?" *Biological Invasions* 13 (2011): 793–802.

88. Dorcas and Willson, *Invasive Pythons in the United States*, 3.

89. Robert N. Reed and Ray W. Snow, "Assessing Risks to Humans from Invasive Burmese Pythons in Everglades National Park, Florida, USA," *Wildlife Society Bulletin* 38 (2014): 366–69.

90. Pasko and Goldberg, "Review of Harvest Incentives to Control Invasive Species," 264.

91. "The Economic Cost of Large Constrictor Snakes," *U.S. Fish and Wildlife Service Bulletin*, January 2012, 1.

92. Dorcas and Willson, *Invasive Pythons in the United States*, 3.

93. Ricardo Betancur et al., "Reconstructing the Lionfish Invasion: Insights into Greater Caribbean Biogeography," *Journal of Biogeography* 38 (2011): 1282.

94. Benjamin I. Ruttenberg et al., "Rapid Invasion of Indo-Pacific Lionfishes in the Florida Keys, USA," *Bulletin of Marine Science* 88 (2012): 1051–59.

95. Thomas K. Frazer et al., "Coping with the Lionfish Invasion: Can Targeted Removals Yield Beneficial Effects?" *Reviews in Fisheries Science* 20 (2012): 185–91; Kristen A. Dahl and William F. Patterson III, "Habitat-Specific Density and Diet of Rapidly Expanding Invasive Red Lionfish Populations in the Northern Gulf of Mexico," *PLoS ONE* 9 (August 2014): 1–13; Theodore S. Switzer et al., "Temporal and Spatial Dynamics of the Lionfish Invasion in the Eastern Gulf of Mexico: Perspectives from a Broadscale Trawl Survey," *Marine and Coastal Fisheries* 7 (2015): 1–8.

96. Marissa F. Nuttall et al., "Lionfish Records within Mesophotic Depth Ranges on Natural Banks in the Northwestern Gulf of Mexico," *BioInvasions Records* 3 (2014): 111–15.

97. Luiz A. Rocha et al., "Invasive Lionfish Preying on Critically Endangered Reef Fish," *Coral Reefs* 34 (2015): 803–6.

98. Mark S. Hoddle, "Restoring Balance: Using Exotic Species to Control Invasive Exotic Species," *Conservation Biology* 18 (February 2004): 38–49.

99. Pasko and Goldberg, "Review of Harvest Incentives to Control Invasive Species," 269.

100. Michael R. Rochford et al., "Molecular Analyses Confirming the Introduction of Nile Crocodiles, *Crocodylus niloticus* Laurenti 1768 (Crocodylidae), in Southern Florida, with an Assessment of Potential for Establishment, Spread, and Impacts," *Herpetological Conservation and Biology* 11, no. 1 (April 2016): 80.

101. Ibid., 83.

102. CrocBITE: The Worldwide Crocodilian Attack Database, www.crocodile-attack .info/data/map (accessed August 27, 2016).

103. Qtd. in Coates, *American Perceptions of Immigrant and Invasive Species*, 1.

104. Elton, *Ecology of Invasions*, 15.

105. Quoted in William Balee, "Historical Ecology: Premises and Postulates," *Advances in Historical Ecology*, ed. William Balee (New York: Columbia University Press, 2013), 17.

106. Warren M. Hern, "Why Are There So Many of Us? Description and Diagnosis of a Planetary Ecopathological Process," *Population and Environment* 12 (September 1990): 9–39; Warren M. Hern, "Is Human Culture Carcinogenic for Uncontrolled Population Growth and Ecological Destruction?" *BioScience* 43 (December 1993): 768–73.

107. Elizabeth Kolbert, *The Sixth Extinction: An Unnatural History* (New York: Holt, 2014).

108. Daniel Simberloff, "Introduced Species, Homogenizing Biotas and Cultures," in *Linking Ecology and Ethics for a Changing World* (Dordrecht, Netherlands: Springer, 2013), 33–48.

109. Low, *Feral Future*, xxvi.

110. Jeffrey M. Diez et al., "Will Extreme Climatic Events Facilitate Biological Invasions?" *Frontiers in Ecology and the Environment* 10 (June 2012): 249–57; Jessica J. Hellmann et al., "Five Potential Consequences of Climate Change for Invasive Species," *Conservation Biology* 22 (2008): 534–43.

111. Elton, *Ecology of Invasions*, 159.

POLITICAL-ECOLOGICAL EMERGENCE OF SPACE AND VULNERABILITY IN THE LOWER NINTH WARD, NEW ORLEANS

ROBERTO E. BARRIOS

Of all the New Orleans neighborhoods devastated by Hurricane Katrina's flooding, the Lower Ninth Ward became iconic of the disaster.[1] What brought this part of the city to national attention was the dramatic breach in multiple locations of a human-made levee that allowed massive amounts of water to flow from the Inner Harbor Navigation Canal—better known to New Orleanians as the Industrial Canal—into the neighborhood. This canal was originally built by the U.S. Army Corps of Engineers to expedite the movement of commercial cargo ships between Lake Pontchartrain and the Mississippi River and to provide additional harbor space for the Port of New Orleans.[2]

The sudden rupture of the Industrial Canal's levees provided a path for a powerful torrent that washed away houses and human lives in a matter of minutes. The visual impression created by this destruction caught the attention of news media and turned the eyes of an entire nation to a neighborhood that had a long history of neglect by the city government and of stigmatization by city residents associating poverty, criminality, and violence with working-class African Americans (95 percent of the area's 19,515 residents self-identified as black in the 2000 Census), and not the sociopolitical processes that historically created stark socioeconomic inequities in New Orleans.[3] As resident and community organizer Victoria Jackson told me in 2008, "The city always looked at us as a downtrodden neighborhood."

The newly attained iconic status of the Lower Ninth Ward presents many of its longtime residents with a seemingly contradictory situation. Katrina evinces how human practices can enhance the destructive and socially disruptive capacities of geophysical phenomena; how they can distribute the effects of catastrophe along the lines of gender, age, race, class, and ethnic differentiation; and how these practices manifest spatially (who lives/works/dies where, under what conditions of risk, and bearing what impacts of the disaster).[4] Consequently, the Lower Ninth has become a deeply politicized space where city, state, and national politicians regularly make campaign stops to offer promises of social transformation and environmental justice.

But even as the Lower Ninth has become a nationwide political commodity, the area remains low on the list of city and state government reconstruction priorities. Lower Ninth residents, for example, have systematically received lower compensation for the damages sustained by their houses from the Louisiana Recovery Authority's (LRA) Road Home Program, which has used pre-Katrina property values to estimate homeowners' losses.[5] These pre-Katrina figures, in turn, were the product of spatially arranged racist and classist prejudices that suppressed property values in predominantly African American New Orleans neighborhoods, and their use by the LRA amounted to the perpetuation of racially based inequalities. The Lower Ninth Ward has also continued to exist in a marginal space in relation to city government. In 2008, the Office of Recovery and Development Administration, created by Mayor C. Ray Nagin and directed by Edward Blakely, scheduled twenty-two recovery projects for the minimally damaged areas of Uptown and Audubon (where the city's premier elite neighborhoods are located), one of which included the construction of tennis courts, while the Lower Ninth Ward was scheduled to receive only three. It is noteworthy that the combined populations of Uptown and Audubon before Katrina was comparable to that of the Lower Ninth Ward (21,510 in the former two versus 19,515 in the latter). Furthermore, the Lower Ninth Ward sustained catastrophic damage and flooding exceeding fourteen feet of water, while flooding in Uptown varied from none for households on prestigious St. Charles Avenue to under eight feet in its most affected areas.[6]

For a number of social scientists and theorists, the Lower Ninth Ward could be said to provide a prime example of how space is socially produced and how its production instills and sustains inequities and disaster vulnerability.[7] In this essay, I use the case of the Lower Ninth

Ward to explore the utility and limitations of social-production-of-space approaches in disaster research. Specifically, I make the case that, in light of the reviewed evidence, theories of the social production of space stand to be enhanced by science and technology studies that recognize material agency's role in the human-environment relations that shape catastrophes.[8]

Analyses of disasters that approach space (and vulnerability, for that matter) as a social product, I argue, are limited by their sole focus on the role of human actors in the shaping of territories where disasters take form and magnitude. A case in point is southeastern Louisiana, with its three-hundred-year history of colonial and capitalist modification of the environment. Granted, human alteration of the lower Mississippi Delta occurred prior to European colonization in the form of relatively limited modification of riverbanks and the construction of midden mounds, but as a number of social scientists have demonstrated, the practices that shaped Hurricane Katrina's destruction were the extensive levee construction, river and port engineering, wetland degradation, and urbanization that characterized the region following its eighteenth-century colonization.[9]

In an approach based on the social production of space, people's actions—such as levee and canal building, the construction of urban areas, and wetland destruction—take center stage when analyzing the origin of disasters. Recent advances in science and technology studies, however, have noted the role of material agency in such processes; that is, the ways environmental features and their reactions to human action are not solely within the control of human beings and manifest in unexpected ways in relation to human practice.[10] Consequently, I propose a shift in the theorization of space and disaster vulnerability from something that is strictly socially produced to something that is politically-ecologically emergent. My emphasis on political-ecological emergence is meant to draw upon the co-constitutive and ever-shifting relationships between people (with their values, policies, technologies, politics, and social orders) and the material world that analyses of science and technology studies strive for.[11] At the same time, I show how a political-ecological approach to the emergence of space helps us understand why many residents of the Lower Ninth Ward continue to experience the paradox of simultaneous disaster iconicity and sociopolitical marginalization in the aftermath of Hurricane Katrina and how this situation may be addressed.

Theorizing the Spatiality of Vulnerability and Disasters: From Social Production to Political-Ecological Emergence

Critical social theorists are unique in their approach to the study of space in that they do not accept it as something that precedes social action and meaning-making.[12] Instead, there are long-established currents of thought in the social sciences and humanities that insist that space must be understood in relation to the people by whom it is represented, structured, and experienced.[13] A key text in this line of inquiry is Henri Lefebvre's *The Production of Space,* in which he proposed "a unitary theory" that recognized space as a social product characterized by three distinct but interrelated "moments": the lived, the conceived, and the perceived.[14]

In Lefebvre's model, lived space corresponded to the ways space is organized and structured as a result of a population's distinct practices of commodity production, circulation, consumption, and biological reproduction. This included the layout of infrastructure like roads, canals, and agricultural lands; the construction of factories or workshops; and the particular construction and design of housing structures and leisure spaces.

Conceived space, or space of representation, referred to the ways politically influential groups conceptualized space in a given social context or influenced the conceptualization of space in urban planning and architectural design. Finally, perceived space, or representational space, referred to the ways people experienced and remembered space. Within this moment of space, one could include memories, rituals, and symbolic practices through which people make space meaningful in everyday practice. Together, the relationships between these three moments, Lefebvre argued, granted social scientists an understanding of the connections between the ways people think about, experience, and do things with/in space in a particular social context.

The utility and limitations of Lefebvre's model may be explored by looking at the spatial history of the Lower Ninth Ward before and after Hurricane Katrina. By spatial history, I mean the answers to the following questions: Who built what, at what time, and for what reason? How was the built environment experienced, and by whom? What kinds of practices took place in this built environment, and in what ways did they further or subvert the agendas of its creators? What meanings did this built environment take for residents of New Orleans? What kinds of persons with what kinds of sensibilities came into being over the course of life experiences of these spaces? Finally, how have these relationships between space and personhood manifested over the course of post-Katrina reconstruction, and how have they mattered in the interactions between residents of the Lower Ninth Ward and city government officials, nonprofit program managers, environmentalists, activists, and academics that comprise the recovery process?

At the same time, I want to show how the case of the Lower Ninth Ward invites us to reconsider Lefebvre's idea of space as something that is strictly socially produced and to consider, instead, space as something that is politically-ecologically emergent. By this I mean to add yet another question to the preceding line of queries: What role, if any, did material agents like water salinity levels, rivers, cypress swamps, and navigation canals play in these processes of meaning-, person-, and vulnerability-making that characterize spatial emergence? A key issue in this reconsideration is the way Lefebvre conceptualized "nature," or "physical space," as something that is passive, without agency, and subject to domination by human beings. Writing about nature's role in the social production of space, Lefebvre noted:

The first implication is that (physical) natural space is disappearing. Granted, natural space was—and it remains—the common point of departure: the origin and the original model, of the social process—perhaps even the basis of all "originality." Granted, too, that natural space has not vanished purely and simply from the scene. It is still the background of the picture; as decor, and more than decor, it persists everywhere, and every natural detail, every natural object is valued even more as it takes on symbolic weight. . . . Yet at the same time everything conspires to harm it. The fact is that natural space will soon be lost to view. . . . True, nature is resistant, and infinite in its depth, but it has been defeated, and now waits only for its ultimate voidance and destruction.[15]

The problem with Lefebvre's model of social production is that it represents "nature" as a passive object that stands opposite to social practice. In this analytical approach, agency lies completely among people, while "nature" is simply an inert object to be dominated and destroyed (or perhaps, simply to serve as a background). In recent years, anthropological and sociological studies of science and technology, political ecology, and disaster vulnerability have argued that such a representation of nature is a product of modernist thinking that falls short of apprehending how human-environmental relationships actually work.[16]

Bruno Latour, for example, has shown how the idea of social production reduces our understanding of complex phenomena involving human and material agents strictly to relationships among (and the actions of) people, leaving out the role of materiality (whether human-created objects like technological devices or environmental features like rivers and soil) in the shaping of human values, subjectivities, and political ecologies.[17] Timothy Mitchell and Andrew Pickering, in turn, have argued that modernist perspectives on the environment have routinely featured the representation of salient landscape features such as the Nile and Mississippi rivers as passive and preexisting elements of nature when, in fact, these rivers have long histories of modification through human actions (for example, the construction of levees that fix a river's path and limit its sedimentation processes).[18] At the same time, these bodies of water have also acted in ways that subverted people's intentions, manifesting agency that "pushed back" on human practices and influenced social values.[19] Finally, Tim Ingold has made the point that ecological studies that see the natural environment as something that predates the entry of people are not at all ecological. A true ecological approach is one that understands environment and people as historically co-constituting entities that never existed without one another and cannot precede the other's entry, as such actions presuppose the independent existence of both.[20]

For the above-named scholars, "nature" is by no means passive or clearly separable from "the social," and they envision the researcher's task as understanding how people and environments mutually shape one another in the moment of practice. By using the term "emergence" rather than "production," I intend to overcome the anthropocentric emphasis of Lefebvre's theory of spatial production where it is humans and only humans who play a role in the making of space, and shift the focus to the co-constitutive human-material relations of political ecology—political ecology, in this case, being an approach that recognizes the dialectical relationships between policy, environment, social practice, and the spatial manifestations of capitalist production and circulation.[21]

Over the course of the last three hundred years, New Orleans residents have engaged in practices that helped shape the city's various spaces and, in the process, have created landscapes of social differentiation, inequity, and varying forms of vulnerability to hazards including flooding and communicable diseases, as Greg O'Brien's and Urmi Engineer Willoughby's essays in this volume demonstrate. These practices have also inscribed the history of economic liberalism through its various phases (from the very first experiments with capitalist investment and paper money in the early eighteenth century to neoliberal imaginings of the city's reconstruction in post-Katrina recovery planning and practice) in the space(s) of Greater New Orleans. As in Christopher Church's analysis of the ways in which the creation of a sugar-plantation economy in South Florida dramatically transformed the region's landscape and inequitably distributed flood risk along lines of class and race during the hurricane of 1928 (also in this volume), the history of capitalism in the Mississippi Delta has similarly shaped both the landscape and the disaster vulnerability of the region. People's space-making practices, however, are not the whole story. Nonhuman agents (including environmental features like rivers, gulfs, saltwater, cypress forests, and bayous) have also played significant roles in the shaping of the city's emergent spaces, their meanings, and their vulnerabilities.

Critical readers may rightly question whether the approach I am proposing verges on ascribing a quasi-religious quality to the material world and therefore reinstating the trope of a vengeful and punitive "mother nature" in disaster studies. I emphasize that I am not proposing an animistic or anthropomorphic understanding of the environment's agency, nor am I attributing sentience or intentionality to the landscape. Instead, I conceptualize agency of things like rivers and saltwater as co-constitutive and relational, that is, their capacities manifest in relation to other things and people.

Readers may also wonder why I choose to refer to the ways bodies of water and landscapes react to human actions as examples of material agency and not simply "unexpected consequences of development projects." The destruction of wetlands and enhanced flood risk that resulted from the post-eighteenth-century development of the lower Mississippi can certainly be thought of as "unexpected consequences," but doing so limits our analysis of who and what has agency in the process of space production. In order to innovate theoretically, it is necessary to deal with these unexpected consequences in terms of agency.

The Political-Ecological Emergence of the Lower Ninth Ward

Before European colonization, the Mississippi River was a collection of streams and oxbow lakes that were not always continuous and whose path varied over the course of the region's geological history. The river as we know it today, which is an uninterrupted and navigable structure that facilitates the transportation of mass-produced commodities, is partly the product of human practices (such as levee construction or river engineering) intended to create a commercial channel.[22] While the Mississippi was an adequate passageway for European watercraft in the 1700s, large ships were still subject to changing tides and the complicated navigation of sandbars. Early eighteenth-century levee construction was limited to the areas surrounding settlements and plantations for the purposes of flood protection for crops and people. By the nineteenth century, levee construction shifted to the creation of a continuous water channel that facilitated the movement of people and commodities from the Gulf of Mexico to what is today's U.S. Midwest.[23]

At the same time, not all of the Mississippi's qualities are solely human-made. The leveeing of the river, which prevented it from doing those things rivers normally do (meander, occasionally shift course, carry and deposit sediment, and escape its banks), has led to unexpected manifestations of material agency. Such is the case of the elevation of the river over New Orleans, as sediments that can no longer find their way into the river's floodplain due to human-made levees have, instead, settled on the river's bottom.[24] This process has resulted in the continued rise of the river above New Orleans, increasing the possibility of catastrophic flooding during events like Hurricane Katrina. Because this sedimentation was once also responsible for the creation of the delta, its constraint by artificial levee systems means the coastline is no longer replenished. Furthermore, subsidence and coastal erosion caused by canal-building practices of the regional oil industry (which lead to saltwater intrusion that destroys land-building vegetation) also enhanced the exposure of the city to disaster-triggering hazards like storm-tide surges.[25]

It is worth noting that New Orleans was not flooded by water from the Mississippi River during Hurricane Katrina. Instead, the city was inundated by multiple drainage-levee failures that pumped sewer water out to the adjacent Lake Pontchartrain and the breaches of the Industrial Canal, which was filled to capacity by storm surge from the Gulf of Mexico. Nevertheless, the Mississippi River's flood hazard remains a key concern of city residents and flood engineers alike.

Along this same river, just east of New Orleans's eighteenth-century city limits, lies the area of the Lower Ninth Ward, which gets its name from voting districts first delimited in 1809. The Ninth Ward itself was not officially recognized until 1852, when its boundaries consisted of Almonaster Avenue, Lake Pontchartrain, the St. Bernard Parish boundary, and the Mississippi River. Over the course of the 1800s, people transformed the area from plantations to small family homes and farms. The majority of these New Orleanians were recently arrived immigrants from Ireland, Germany, and Italy as well as African Americans emancipated from slavery in the 1860s.[26] Between 1918 and 1923, the construction of the Industrial Canal divided the Ninth Ward in two, creating the upper and lower sections (upper indicating "upriver" and lower "downriver" location) from which the Lower Ninth Ward gets its name.[27] The intention behind the canal's construction was the provision of a shortcut for commercial ships from Lake Pontchartrain to the Mississippi River and the expansion of the Port of New Orleans's wharf space.

The Industrial Canal provided a great financial benefit to the Port of New Orleans, cutting costs by expediting travel and increasing profits by providing expanded wharf space, but its construction also enhanced the neighborhood's flood risk by bringing lake and river water into the area. The canal also isolated the neighborhood from the city's economic centers, the French Quarter and Central Business District. From Lefebvre's perspective, the construction of the canal corresponds to the spatial practice of capitalism, in which people create the space of the canal with the intention of expediting the circulation of mass-produced commodities. The divided spaces created by this canal also took distinct meanings over the course of the city's history, with the Lower Ninth being seen by many nonresidents as a marginal neighborhood even as it became a representational space for its residents: a place of ritual-, memory-, and personhood-making.

The political ecology of the Lower Ninth Ward was further complicated in 1958, when the Army Corps of Engineers began the construction of yet another major navigation artery, the Mississippi River Gulf Outlet (MR-GO), which connected the Industrial Canal with the Gulf of Mexico. The canal had a number of unexpected effects on the neighborhood's surrounding wetlands, such as increasing salinity levels and degrading surrounding cypress forests that buffered the Lower Ninth area from tropical storms. The new channel also created another pathway into the neighborhood for storm surges. Similar to the case of the Industrial Canal, the motivation behind the construction of the MR-GO was the speeding up of commercial cargo ship movement from the Port of New Orleans to the Gulf of Mexico. This is, yet again, another spatial practice of capitalist development, where environments are transformed, first and foremost, for the circulation of commodities and labor and the replication of capital.[28]

But the case of the Lower Ninth Ward also shows how these practices dialectically enhanced the agencies of geophysical phenomena (salinity levels, hurricanes) in unexpected ways, leading to unexpected conditions of enhanced flooding risk. The MR-GO, which engineers intended to remain a channel no more than three hundred feet in width, widened up to one and a half miles in some parts as saltwater from the Gulf of Mexico entered the canal's adjacent bayous and cypress forests, destroying vegetation and turning marshes into open water (an environmental impact that was not initially predicted by the U.S. Army Corps of Engineers). Unlike Lefebvre's unitary theory of space, in which "nature" is merely destroyed and has no agency in the production of space, agents like saltwater also played a role (even if an unanticipated one) in the emergence of vulnerable spaces in southeastern Louisiana.

During Hurricane Katrina, the MR-GO and Industrial Canal functioned together in ways undesired by river engineers to devastate the Lower Ninth Ward. The MR-GO provided a channel cleared of flood-protecting cypress swamps for storm surge to enter the area, while the Industrial Canal provided a space for Gulf water to enter and under-maintained levee structures to fail in multiple localities. The unexpected and unpredictable responses of material agency to the environmental modifications of Corps of Engineers channeling projects are another example of how technology often has unintended consequences, and these consequences are prime examples of what Andrew Pickering and Joseph Masco have identified as the dialectically emergent and never fully predictable quality of material agency. To further complicate the emergent and unpredictable vulnerabilities engendered by Corps of Engineers navigation projects, the politics of flood-protection-system revenue generation and project design outlined by Kevin Fox Gotham (in this volume) further compounded the area's propensity for catastrophic flooding.

Space, Identity, and Inequity in New Orleans

Processes of spatial emergence, however, are not limited to river engineering projects. They also involve human-environment relationships where people give meanings and forge racialized, gendered, and class-based identities vis-à-vis emergent spaces. Additionally, it is the interplay between development practice, spatial emergence, vulnerability enhancement, and identity formation that generates and sustains the contradictions of disaster recovery confronted by Lower Ninth residents. It is for this reason that I now turn to the history of race and class relations in the Lower Ninth.

The Lower Ninth Ward was more socially diverse during its early years than at the turn of the twenty-first century when Hurricane Katrina made landfall. In 1960, New Orleans counted 627,525 residents, of whom 233,514 (37 percent) self-identified as black in the U.S. Census. By 2005, in contrast, New Orleans's total population had declined to 462,269 people, and the percentage of residents who self-identified as black increased to 68 percent.[29] These citywide figures resulted from a complex process of urban-suburban flight that was driven by multiple factors, including racism.

Following the end of state-sanctioned segregation in the 1960s, New Orleans witnessed a movement of residents who self-identified as white to its surrounding suburbs in Jefferson and St. Bernard parishes. Meanwhile, a significant number of middle-class African Americans moved to New Orleans East and Pontchartrain Park. Many African Americans also left Greater New Orleans in search of jobs in states where the labor market was not as rife with racial discrimination.[30]

In terms of Lefebvre's three moments of space, the racial polarization witnessed by many New Orleans neighborhoods, including the Lower Ninth Ward, shows how the spatial practices and lived spaces of New Orleans served as key sites for the making and sustaining of racialized differences among city residents. During segregation, the spaces of social and human reproduction (segregated public schools and housing) sustained the separation of bodies upon which continued racial differences were dependent. In New Orleans, cultural practice has a lengthy history of serving as a mechanism for denoting and embodying racial identity and maintaining racial difference; in other words, how a person behaves is thought to partly denote their racialized identity and therefore index their place in society.[31] The separation of these spaces created distances that helped prevent cultural mimesis (the purposeful or inadvertent imitation of behaviors, linguistic practices, bodily stances, music, and rituals of people deemed "other") that could obfuscate racial lines. Desegregation threatened to undermine these mechanisms of race-making, and as a result, a significant proportion of New Orleanians who self-identified as white devised new spatial practices of social distancing and differentiation through, for example, the construction of suburbs, de facto segregation through police profiling, and the creation of new school districts.[32]

During the early twentieth century, the Lower Ninth Ward was specifically developed as a place where working-class African Americans could own property in the context of state-sanctioned segregation. While channel and levee construction practices shaped this part of Greater New Orleans as a space of capitalist maritime circulation, material agency manifested in unexpected ways, making the area a space of flood vulnerability. Simultaneously, socio-spatial practices of racial differentiation contributed to forming the Ninth Ward as a site of racialized environmental injustice.

The racial motivations behind New Orleans's demographic shift were also nuanced by the meanings people attributed to life in suburbia (the imagined escape from inner-city "problems," and the association of modernity with suburban life) and the ways federal subsidy programs encouraged home ownership in outlying suburban areas.[33] Broader economic issues like the city's financial crisis during the 1980s and the downscaling of oil company operations also played a role in the city's population shift, leading to continued demographic decline.[34] Meanwhile, suburbs that became renowned as places of suburban flight for African Americans such as New Orleans East came to face social and disaster vulnerabilities like such as subsidence caused by wetland drainage and property devaluation.

By 2005, the Lower Ninth Ward had experienced a process of racial polarization. The urban-suburban flight that created this polarization was accompanied by unofficial local government and policing practices meant to limit the spatial and social mobility of working-class African Americans, and, according to my ethnographic interlocutors, the Lower Ninth Ward became stigmatized as a place of poverty and perceived criminality. But the neighborhood was more complex on the eve of Hurricane Katrina than these representations suggested. The general area of the Lower Ninth Ward was divided into two neighborhoods, and residence in these areas became yet another spatial dimension of the identities of the New Orleanians who lived there. The area's southern edge, which extends from St. Claude Avenue to the Mississippi River and features higher elevation thanks to the river's natural levees, is recognized as the Holy Cross neighborhood by residents and city officials, while the area extending northward from St. Claude Avenue to Bayou Bienvenue (a prominent wetland area) is simply referred to as the Lower Nine. In pre-Katrina times, residence in either neighborhood was an important marker of identity.

The pattern of identity formation is yet another instance that demonstrates the role of material agency in the political-ecological emergence of space. The natural levees created by the Mississippi River prior to the city's founding, less prone to flooding, have historically been associated with the residences of local elites. On the other hand, the areas extending away from the river's natural levees are characterized by lower elevations and greater flood risk, and have a history of being associated with the homes of servants and working classes.[35] The areas that were originally built up by the river's sedimentation patterns, then, have been interpreted and made meaningful through discourses and practices related to class-formation—actions that are also inflected with ideas about racial difference. While social relations and interpretations are certainly a part of the process through which space manifests in New Orleans, it is also important to account for the role of material agents like rivers and natural levees in the way the city's spatial structuring and flood vulnerability have come into being.

Contested Visions of Neighborhood Well-Being

In the three decades preceding Katrina, city government declared Holy Cross a historic neighborhood, and the area saw an influx of new residents attracted by low property values and historic homes. Holy Cross also witnessed the formation of a neighborhood association that rose to prominence as a grassroots-organizing leader after Hurricane Katrina, and efforts by historical preservationists seem to have played an important role in helping this part of the Lower Ninth Ward recover faster than its northern counterpart.[36] Although the two neighborhoods had broad socioeconomic similarities, many Lower Nine residents felt that some of their Holy Cross counterparts thought of themselves as distinct from (and perhaps even better than) them.

Post-Katrina, Holy Cross residents accused of elitism feel misunderstood when confronted with the sentiments of their fellow Lower Niners. Nevertheless, some Holy Crossers continue to engage in practices that have long histories of denoting hierarchized and racialized class differences in the city.[37] These practices include the avoidance of or disdain for neighborhood working-class bars, the gentrification of historic neighborhoods, and the monitoring of working-class residents' use of street spaces for daily socialization. As Setha Low has commented, neighborhood associations have a tendency to articulate subtle forms of spatial segregation through the seemingly harmless act of urban beautification.[38]

Space, Identity, and Politics after Katrina

In the disaster's aftermath, people in the Lower Ninth Ward have continued to struggle with the subtle differences in spatial practice and neighborhood identity that differentiated (and were part of the spatial emergence of) the Lower Nine from Holy Cross before the storm. These struggles stem from the sentiment of some Lower Nine residents that Holy Crossers have customarily looked after their own self-interest instead of the well-being of the Lower Ninth Ward as a whole, and that they have often secured resources from city government and philanthropic organizations without sharing these with other neighborhood organizations across St. Claude Avenue—hence reinscribing a spatial difference. These tensions among Lower Ninth Ward residents present a significant challenge to all neighborhood organizers, regardless of which neighborhood identities they sympathize with, as it is their shared sentiment that a broad-based constituency is more efficacious at making demands from local government officials for reconstruction resources.

Over the course of my ethnographic research, I also learned that resident organizers of various walks of life are aware of the ways city government officials (New Orleans City Council members in particular) exploit the neighborhood's diversity, insisting that reconstruction projects are not allocated to the neighborhood because of residents' "incapacity" to speak with one voice—reasoning that denies the historic stigmatization and marginalization of the Lower Ninth Ward in city politics.

Post-Katrina, neighborhood organizers disagree on how differences in spatialized identities, spatial practices, and visions of recovery can be negotiated. Some Holy Cross neighborhood organizers insist on "looking forward" or "looking to the future" instead of "harping on past differences" as a means of building a cohesive neighborhood constituency. But other residents interpret these calls to forget the past and look to the future as attempts to dismiss what remain serious concerns in the neighborhood (most notably, gentrification and elitism). The latter residents, in turn, emphasize the need to look to the past for important insights from which to devise new discourses of neighborhood identity (and practices of sustainable recovery) that overcome local spatial and social divides.

Issues of neighborhood identity and spatial emergence came to the fore during a discussion among neighborhood leaders from Holy Cross and the Lower Nine in June 2008. The organizers held a day-long event during which they discussed strategies for better coordination among neighborhood groups. As they discussed this topic, John Jackson, a resident-organizer from Holy Cross, shared his assessment of neighborhood identity politics with the dozen or so other grassroots leaders at the table: "Once we rid ourselves of traditional thinking, we can move on to creating the future. I say a lot of times, if you hold on to the past, you can't move on to a brighter future. Not to say you don't look back to the past to get some wisdom. But I know there has been bickering in the Lower Ninth. But Katrina has given us an opportunity to build a better city. I mean, slavery was once a tradition, but we let go of it."

John's comment elicited a response from Bernie Holmes, a respected Lower Nine resident, who disagreed with him:

I am a traditionalist, because I look to the past to see if they have any traditions to make the future better. I think we need to look at the past to see what is going on. Because games are still being played. The money that is coming to our city, how is it being misused? This city is getting millions and millions of dollars, but we don't get it. The majority of the money this city is getting is on the back of the Lower Ninth Ward. That's where the collaboration has to happen.... [Before Katrina] the Lower Ninth Ward provided its own economy; when my dad lived there, a lot of people who lived in the Ninth Ward survived on the Ninth Ward, they didn't have to leave. Now I don't know what's being planned, but they're trying to take it away. You talk about segregation, we have segregation within the Lower Ninth Ward community. It may not be black and white, but it's there. I thought the idea behind this alliance was that we would bring it all back, but that's not happening.

In this exchange, Mr. Holmes insisted that some Holy Cross residents have a history of thinking of themselves as different from and even superior to their Lower Nine counterparts: "*You talk about segregation, we have segregation within the Lower Ninth Ward community. It may not be black and white, but it's there.*" Moreover, Mr. Holmes noted that these identity politics (which are matched by daily practices related to spatial emergence) are still manifesting in the recovery process, leading to an uneven reconstruction of north and south sides of the Lower Ninth Ward.

Part of the challenge confronted by Lower Ninth residents is that Holy Cross organizers often feel unfairly accused of intentionally creating hierarchies and inequities within the neighborhood when, from their perspective, all they are doing is trying to enhance quality of life in the area. Nevertheless, the actions of some Holy Cross residents, whether intentional or not, do create socioeconomic distinctions that manifest spatially. The principal neighborhood association in Holy Cross, for example, has a lengthy affiliation with New Orleans historic preservation groups that restore and sell homes at values that significantly surpass those of comparable properties in the area. Renovated historic homes tend to be valued between $140,000 and $230,000. By contrast, Jeanelle Holmes, a lifelong resident of the Lower Ninth Ward who became one of my key interlocutors over the course of this ethnographic study, purchased her historic house on Flood Street for $20,000 and refurbished it with a $60,000 bank loan. Jeanelle's modifications to the property did not meet historic restoration standards, and doing so would have greatly increased her costs. Instead, her renovations kept a historic property from reaching a state of complete disrepair and made it usable for a lifelong neighborhood resident. Practices such as historic preservation, then, enact subtle forms of spatial exclusion, raising property values, attracting new homeowners who do not have a history of residence in the neighborhood, and pushing away New Orleanians of modest means.

An approach to the Lower Ninth Ward that understands the neighborhood's space as politically-ecologically emergent helps us recognize that practices of space-making and spatial distancing have a long history as mechanisms of differentiation and exclusion in New Orleans, and that the Lower Ninth Ward has not remained untouched by these trends. Rather than denying the presence of spatial hierarchies and hierarchized identities, an approach to neighborhood recovery that focuses on spatial emergence can help neighborhood organizers recognize the role of their own actions (in addition to those of politically powerful agents like city council members, mayors, real estate developers, and other New Orleanians) in the making of difference and inequity—a necessary step for broad constituency-building.

Socio-Spatial Complexity, Environmentalism, and Sustainability

The recovery of the Lower Ninth Ward is taking place within a broader context of neoliberalization of disaster reconstruction, a movement in which government officials and policy makers see market liberalization, the privatization of public services and resources, and the subjection of all aspects of social life to capitalist cost-benefit analysis as necessary practices for producing well-being.[39] As Ted Steinberg has argued, neoliberalism is the ideology that free-market forces and individual choice are the most effective mechanisms for addressing the challenges of living in collective societies, even environmental degradation and sustainability.[40] In the case of New Orleans's post-Katrina recovery, local and federal government officials have also fashioned neoliberalism as a form of corporatism where recovery resources are funneled from state agencies to for-profit companies. This has been done either through the use of public funds to encourage private state investment, or through the contracting of for-profit companies to handle case management and aid distribution. For example, the city government's Office of Recovery and Development Administration's recovery plan focused on the creation of seventeen target zones. The plan proposed the use of eminent-domain laws and reconstruction funds to create nodes of investment (such as shopping malls and film studios) that would one day produce tax revenue to fund the public services many New Orleanians urgently need.[41] In the meantime, critical public services like the city's Charity Hospital, which was a regional health resource for economically marginalized Louisianans, remained without replacement ten years after the storm.[42]

The neoliberal approach to reconstruction has created a gap in the provision of public services and mitigation efforts in the Lower Ninth Ward, which a number of academics, environmentalists, conscious capitalists, and nonprofits are attempting to fill. All of these organizations and experts are interested in contributing to the neighborhood's recovery, but they are also driven by their own agendas and visions of well-being and sustainability. The degree to which they have been able to negotiate their priorities with neighborhood residents—who are themselves a heterogeneous group with differing visions of neighborhood well-being—varies.

Some residents, for example, sympathize or identify with the neighborhood's predominantly African American working class and working-poor population and advocate for recovery practices that prioritize the reinstatement of the neighborhood's pre-Katrina population. Other residents collaborate with historical preservation organizations whose projects raise property values and encourage the introduction of new upwardly mobile people into the Ninth Ward. Over the course of post-Katrina recovery, these different takes on neighborhood development have become the subject of heated debate, further complicated by the introduction of environmental sustainability agendas (which meant to address the impact of capitalist spatial practices) by nonprofits and academic researchers.

The socio-spatial complexity of the Lower Ninth Ward has a number of implications for environmentalist projects spearheaded by academics and graduate students from state universities and external nonprofit organizations. In Holy Cross, a neighborhood association has developed a center devoted to sustainable reconstruction, and this center has become a key site of engagement between academics, external philanthropic organizations, and a select group of residents. In the aftermath of the storm, discourses of environmentally sustainable reconstruction have become one of the principal means of envisioning the area's recovery, but these discourses have the potential to be conjoined with those pre-disaster practices many Lower Niners interpreted as intra-neighborhood elitism and segregation.

To counter this tendency, some Lower Niners are actively attempting to transform environmentalist discourses and projects, and to invent new ways of "being green" that are sensitive to the particularities of their two neighborhoods. In terms of spatial emergence, discussions about the restoration of adjacent wetlands—which became spaces of devastation through the interplay of spatial capitalist practices of maritime circulation and the unexpected manifestation of material agency (Bayou Bienvenue, in particular)—highlight another important theme: the relationships between what Henri Lefebvre called representational spaces, the agency of environmental features, and the shaping of personhood.

For some prominent resident organizers, wetland restoration projects spearheaded by academics and researchers focus on definitions of environment that emphasize material indicators like water salinity levels and cypress forest density, and downplay what they consider should be the area's central reconstruction priority: the return of displaced neighbors and the restoration of the area's social landscape. Meanwhile, graduate students and researchers from universities involved in these projects respond that their programs prioritize community involvement in wetland restoration, but their notions of community involvement figure Lower Ninth Ward residents as mainly supporters of wetland restoration programs and not as their focal concern.

These different perspectives on what Bayou Bienvenue *is,* how people relate to it, and how its restoration should proceed are manifested spatially in the projects of research universities working in the neighborhood. In terms of spatial theory, one could say that university-driven environmental restoration programs focus too closely on "environment" as a space of nature that is separate from people, and relate to Lower Ninth Ward residents only in their potential role as supporters of its restoration. Many Lower Ninth Ward residents, on the other hand, speak of the bayou as a space where they came into being via formative experiences, and see their displaced pre-Katrina neighbors as an indispensable part of the bayou's space that must be restored just as much as cypress forests and freshwater.

In 2008, for example, students from a major research university constructed an observation deck as part of their bayou restoration efforts. The boardwalk was intended to serve as an attraction that encouraged neighborhood residents' interest and participation in restoration projects. Over the course of neighborhood association meetings, students involved in the project routinely invited Lower Niners to visit the observation deck. Nevertheless, their own neighborhood surveys demonstrated low rates of resident use. The deck, in Lefebvre's terms, is part of a space of representation: the ways hegemonic actors in a given society think about and encourage others to perceive space. In this case, the bayou is conceptualized as an object to be observed from the boardwalk, a spatial distribution of people and "nature" that keeps these two entities separate even as its conceivers and makers claim to bring them together. Furthermore, some prominent grassroots organizers routinely saw such actions as sidelining the restoration of the area's human landscape.

These issues surfaced during a meeting called by grassroots organizers in fall 2009 that brought together faculty from Louisiana State University, Colorado State University, representatives of affordable- and green-energy nonprofits, and neighborhood residents. The meeting was held at a warehouse that Ward "Mack" McClendon, a lifelong resident of the area, converted into a community center.

Before Katrina, Mack dedicated himself to restoring classic cars and driving tow trucks. After the storm, he felt a call to become a grassroots organizer in light of the slow recovery of the Lower Nine and the low rate of return of his pre-Katrina neighbors. Mack's investment in this warehouse contrasted significantly with the student-built boardwalk, materializing the differences at stake in the recovery priorities of community insiders and outsiders. Following a presentation about the bayou's restoration by wetland scientists, Mack commented, "It used to be just like you're talking about, back there [Bayou Bienvenue], I think if we get that back again, it would be good, and the people, getting back to the people, they totally understand that, given the opportunity, specially the elderly. 65 percent of this property was owned by elderly, they're less than 5 percent back, so I think the key is: Don't put the wagon before the horse."

By warning landscape architects and wetland ecologists to not "put the wagon before the horse," Mack made the point that the approaches to wetland restoration he had seen in the neighborhood thus far featured conceptualizations of sustainable reconstruction that defined the region's ecology in too-narrow terms. These were terms that defined the bayou's restoration as the achievement of specific salinity levels and forest density. For Mack and other Lower Nine residents, such conceptualizations of ecology created the sentiment that the restoration of the neighborhood's pre-Katrina population (and their accompanying social relations) was effectively made a secondary priority to "environment" as defined in terms of wetland science.

Alternative discourses of environment and ecology articulated by many Lower Nine residents represent the bayou as a gendered space of memories and rites of passage that could not be thought of in strict wetland-science terms, what Lefebvre would call a representational space. In these alternative discourses, the bayou is figured as a place of childhood adventure where male residents fished, hunted, and made friendships. In the following interview excerpt, Mr. Smith, an elderly Lower Nine resident, recounts his experiences in the bayou using the pronoun "we," indicating that his experiences were always shared with other neighborhood boys:

I remember we used to go back on these tracks, we used to call it Florida Walk, and we used to go back there, sometimes we'd catch a train, go one way, and catch a train and go back the other way, there was a lot of trains because there was a lot of traffic out there on the trains. But I remember I used to go back there when I was young and go fishing, we'd catch snakes. . . . They had all kinda animals back there, and they had a guy that was further down around Southern Scrap, he had a house out there, and he had a stream that used to go by his house and he used to rent you boats and we used to go out and fish and all of that and come back . . . and some people don't believe this here, I remember they had horses back there, at Southern Scrap! Wild horses! I tell people that and they don't believe me! They had wild horses back there. Bring you back. This is in the early fifties or middle fifties.

Mr. Smith's comments also bring to mind Lefebvre's observation that, "if there is production of the city, and social relations in the city, it is a production and reproduction of human beings by human beings, rather than a production of objects."[43] I have made the case that Lefebvre's quote needs to be modified to account for the role of nonhuman entities like Bayou Bienvenue in the production of humans and their forms of personhood. Academic and environmentalist approaches to the bayou's restoration that focus too closely on wetland science and the bayou as a "scientific object" miss this important insight.

For residents of the Lower Nine, the bayou is a place of memories that are intimately linked to the social relations they have experienced over the course of their lives in the area, and these social relations are as much a part of what the bayou *is* as salinity levels and cypress trees. This means that the integration of community into wetland restoration projects must involve more than the recruitment of resident support for environmentalist projects in the Lower Ninth. Instead, resident return and the reinstatement of the pre-Katrina social fabric must be a key priority of wetland space restoration agendas in this New Orleans neighborhood.

Over the course of my ethnographic experiences in the Lower Ninth Ward, I routinely heard comments from other residents that echoed Mack's sentiments. On one occasion, neighborhood grassroots historian and museum curator Ronald W. Lewis addressed the staff of Make It Right, a nonprofit managing the construction of energy-efficient housing. Make It Right managers had arranged a discussion between landscape architects and Lower Nine residents about the geographic integration of the neighborhood with the region's wetland features, including the Mississippi River and Lake Pontchartrain. Responding to a landscape architect's comments that focused on the geophysical features of Greater New Orleans but excluded a discussion of the neighborhood's human and social landscape, Mr. Lewis commented, "The people who built this neighborhood worked the sugarcane fields. They bought lots for 200 dollars and built it themselves. They did it without architects. We didn't have a lot of amenities before. Put our people back, put our tax base back. Without the people we are nothing."

Like Mack, Mr. Lewis insisted that discussions of environmental restoration in the Lower Nine should not sideline discussions of population return, as people and environment are inseparably intertwined and co-constituted in the process of spatial emergence: *"Without the people, we are nothing."*

On July 27, 2013, I caught up with Mack once again, to get his perspective on the recovery of the Lower Ninth Ward over the preceding years. We met in the garage of a house he rehabilitated to host volunteers who were still contributing to the neighborhood's reconstruction. Mack sat on a lawn chair by a small coffee table where some dominos, cigarettes, and other miscellaneous items rested. Mack, known for his unending optimism, was not happy with the way things were going. Property values had increased along with people's homeowner taxes, and he felt the poorest and oldest residents of the Lower Ninth Ward would not be able to afford living there in the near future. The spatial practices of disaster recovery in the Lower Ninth Ward, it seemed, were only prolonging the disaster for the most vulnerable: "In ten years, only a very small minority of the residents will be pre-Katrina Lower 9'ers. Next month, will be eight years. We got less than one third of my community back. That is not acceptable. This community had the highest rate of homeownership in the state. We are treating them like they broke the levees and we gave them a one-way ticket out. The Lower 9th Ward was the hardest hit and the slowest to come back. I think it hasn't been done because of greed. In ten years, it's going to be 5 percent of the people of this neighborhood. When you talk about gentrification ... I can't imagine another community in the nation where this would happen. It's inhumane you are allowing this to happen. We had seven schools before Katrina, we have one now."

Mack went on to make connections between the community's slow return and the novel kinds of space that emerge as a result of neoliberal recovery. Questioning the neoliberal logic of disaster reconstruction, in which tax revenue will one day—in an indeterminate future—allegedly provide the trickle-down economics necessary to replace damaged infrastructure, Mack commented: "Our public officials say: we can't build the schools if the children are not back, but how can the children return if there are no schools? Our children are suffering because of this. They are dying because they can't come back." At the same time, Mack saw environmental restoration and green rebuilding projects as complicit with this tendency, as they sidelined area residents in the distribution of recovery funds and as a main focus of reconstruction: "In terms of the bayou, they are about to spend millions and billions in that bayou and none of it is going to come here. A certain percentage of the community reconstruction should be done by the people of the community. We're getting 1 percent. How does that empower the community?"

The following two years were particularly difficult for Mack. His daughter died due to complications with a pregnancy, and he lost his home to a foreclosure. He would himself pass away on February 14, 2015, after nine years of working incessantly for the recovery of the Ninth Ward's social landscape.

Recently, Gastón Gordillo has made the case that historic and environmental preservation projects that reflect bourgeois sensibilities toward built and "natural" environments have a tendency to evoke feelings of resentment or indifference on the part of subaltern populations who feel such sensibilities are given preference over their own social lives.[44] In this way, elite or hegemonic ways of relating to and making space have the propensity to enact processes that are socially destructive for those most vulnerable to disasters. This is particularly so when socioeconomically marginalized people and their lifeways are pushed out of gentrified historically preserved areas, or when they see their livelihoods and well-being disregarded in environmental projects. The Lower Ninth Ward is a prime example of this tendency in both respects. The vantage point of the political-ecological emergence of space can be brought to bear on this phenomenon of spatial exclusivity through preservation and restoration.

The Lower Ninth as a post-Katrina space of gentrification and exclusion of working-class African Americans emerged as a result of the political-ecological relationships between people, technology, values, identities, and the material world's agency that gave the disaster its form and magnitude. In the post-Katrina moment, the Lower Nine continues to emerge as a space, albeit as one of exclusion and further marginalization fueled by a neoliberal approach to disaster recovery that sees no obligation to directly address the needs of the city's most vulnerable populations.

Conclusion

An analytical approach focused on the political-ecological emergence of space allows us to recognize the human practices and material agencies involved in the making of urban areas, disaster vulnerability, people's identities, and social inequities. From this perspective, we can recognize that the conditions of disaster vulnerability that the residents of the Lower Ninth Ward have experienced were not of their own making, but were the product of policy decisions and development practices unique to the history of economic liberalism in the region. From this vantage point, we can recognize that the mitigation of Katrina's effects in the neighborhood cannot be addressed through a neoliberal approach that figures the area as a space of public disinvestment (as indicated by the low number of recovery projects in the area in comparison to other New Orleans neighborhoods) in order to make it a site of private investment (as demonstrated by Mack's concern about shifting neighborhood demographics and gentrification). In the same way that Ted Steinberg has questioned the merit of proposing neoliberalism as a solution to the environmental destruction caused by capitalist development in the first place, the case of the Lower Ninth Ward similarly challenges the logic of neoliberal disaster recovery. Instead, a concerted effort must be made by regional and city policy makers to address those space-making practices that prioritized commodity and capital circulation over neighborhood flood protection during the twentieth century and both marginalized and stigmatized the Lower Ninth.

In the context of neoliberal disaster reconstruction, nonprofits, academics, and activists take it upon themselves to address issues of environmental sustainability and justice.[45] The case of the Lower Ninth Ward gives us a unique glimpse into the complexities that disaster-affected populations must face when forming these alliances.[46] As the ethnographic evidence shows, disaster-affected localities like the Lower Ninth Ward have complex social landscapes where people differ significantly in terms of habitus, racialized class identities, and notions of what it means to recover sustainably. This complexity is closely related to the broader processes of spatial emergence that made such urban-neighborhood spaces possible in the first place.

In the Lower Ninth, residents insist that environmental recovery projects must emphasize the reinstatement of the neighborhood's social landscape (that is, assist and ensure that pre-Katrina residents return). These residents also emphasize the neighborhood's proud history of self-reliance, and they are guarded as to the increasing role of nonprofit managers and academics in establishing the agenda of disaster recovery. Residents who articulate these positions also express reservations about the social hierarchies and inadvertent elitisms of some established neighborhood associations. In the context of post-Katrina reconstruction, some of these organizations have added environmental sustainability to their pre-disaster focus on historical preservation, but some outspoken residents remain concerned that unresolved class issues that manifest spatially threaten to conflate "rebuilding green" with gentrifying tendencies. Nevertheless, these same residents do not necessarily think environmental sustainability is inherently incompatible with their visions of neighborhood recovery; instead, their position is that issues of social justice (embodied in the restoration of the neighborhood's pre-Katrina social landscape) must be central to discussions of sustainable reconstruction.

NOTES

This essay is in memory of Ward "Mack" McClendon. Unattributed quotations are from the author's field notes, 2008–13.

1. Rachel Breulin and Helen Regis, "Putting the Ninth Ward on the Map: Race, Place, and Transformation in Desire, New Orleans," *American Anthropologist* 108 (2006): 744–50.

2. Richard Campanella, *Geographies of New Orleans: Urban Fabrics before the Storm* (Lafayette: Center for Louisiana Studies, 2006), 6.

3. "Pre-Katrina Data Center Website," Greater New Orleans Community Data Center, www.datacenterresearch.org/pre-katrina/prekatrinasite.html (accessed July 25, 2016).

4. Anthony Oliver-Smith, "What Is a Disaster? Anthropological Perspectives on a Persistent Question," in *The Angry Earth: Disaster in Anthropological Perspective,* ed. Anthony Oliver-Smith and Susanna Hoffman (New York: Routledge, 1999), 22.

5. Cain Burdeau, "Katrina Homeowners Will Share $62M in HUD Settlement," *Houston Chronicle,* July 6, 2011, www.chron.com/news/nation-world/article/Katrina-homeowners-will-share-62M-in-HUD-2077732.php (accessed November 15, 2014).

6. "Pre-Katrina Data Center Website"; Unified New Orleans Planning, *District 4 Recovery Plan* (New Orleans: Unified New Orleans Planning), 1–10.

7. Setha Low, "Claiming Space for an Engaged Anthropology: Spatial Inequality and Social Exclusion," *American Anthropologist* 113 (2011): 389–91; Henri Lefebvre, *The Production of Space* (Oxford, U.K.: Wiley-Blackwell, 1992), 8–9; Piers Blaikie, Terry Cannon, Ian Davis, and Ben Wisner, *At Risk: Natural Hazards, People's Vulnerability, and Disasters* (New York: Routledge, 1994), 3–20.

8. Andrew Pickering, "New Ontologies," in *The Mangle in Practice: Science, Society, and Becoming,* ed. Andrew Pickering and Keith Guzik (Durham, N.C.: Duke University Press, 2008), 40–45.

9. William R. Freudenburg, Robert Gramlin, Shirley Laska, and Kai Erickson, *Catastrophe in the Making: The Engineering of Katrina and the Disasters of Tomorrow* (Washington D.C.: Island Press, 2009), 45–54; Pickering, "New Ontologies," 40–45.

10. Bruno Latour, *Pandora's Hope: Essays on the Reality of Science Studies* (Cambridge, Mass.: Harvard University Press, 1999), 225–30.

11. Aletta Biersack, "Introduction: From the 'New Ecology' to New Ecologies," *American Anthropologist* 101 (1999): 5–10.

12. Lefebvre, *The Production of Space*, 8–9.

13. Claude Raffestin, "Could Foucault Have Revolutionized Geography?" in *Space, Knowledge and Power: Foucault and Geography*, ed. Jeremy Crampton and Stuart Elden (Burlington: Ashgate Publishing Co., 2007), 129–31; Lefebvre, *The Production of Space*, 3–12.

14. Lefebvre, *The Production of Space*, 3–12.

15. Ibid., 30.

16. Biersack, "Introduction: From the 'New Ecology' to New Ecologies," 5–10; Anthony Oliver-Smith, "Theorizing Disasters: Nature, Power and Culture," in *Catastrophe & Culture: The Anthropology of Disaster*, ed. Susanna Hoffman and Anthony Oliver-Smith (Santa Fe, N.M.: School of American Research Press, 2002), 23–30; Timothy Mitchell, *Rule of Experts: Egypt, Techno-Politics, Modernity* (Berkeley: University of California Press, 2002), 19–31.

17. Latour, *Pandora's Hope*, 217–29.

18. Mitchell, *Rule of Experts*, 19–31; Pickering, "New Ontologies," 40–45.

19. Pickering, *The Mangle of Practice*, 9–15.

20. Tim Ingold, *The Perception of the Environment: Essays on Livelihood, Perception, and Skill* (New York: Routledge, 2000), 18–20.

21. Biersack, "Introduction: From the 'New Ecology' to New Ecologies," 5–10.

22. Charles Camillo and Matthew Pearcy, *Upon Their Shoulders: A History of the Mississippi River Commission from Its Inception through the Advent of the Modern Mississippi River and Tributaries Project* (Vicksburg: Mississippi River Commission, 2004), 1–15.

23. Ibid., 1–15; "MRC History," U.S. Army Corps of Engineers, www.mvd.usace.army.mil /About/MississippiRiverCommission(MRC)/History.aspx (accessed January 12, 2016).

24. Pickering, "New Ontologies," 35–40.

25. Julie Koppel Maldonado, Christine Shearer, Robin Bronen, Kristina Peterson, and Heather Lazrus, "The Impact of Climate Change on Tribal Communities in the US: Displacement, Relocation, and Human Rights," *Climate Change* 120 (2015): 601–14.

26. "Pre-Katrina Data Center Web Site."

27. Campanella, *Geographies of New Orleans*, 12.

28. David Harvey, *The Urban Experience* (Baltimore: Johns Hopkins University Press, 1989).

29. Campanella, *Geographies of New Orleans*, 15.

30. Antoinette Jackson, "Diversifying the Dialogue Post-Katrina—Race, Place, and Displacement in New Orleans, U.S.A." *Transforming Anthropology* 19 (2011): 13–16.

31. Jerah Johnson, "Colonial New Orleans: A Fragment of the Eighteenth-Century French Ethos," in *Creole New Orleans: Race and Americanization*, ed. Arnold R. Hirsch and Joseph Logsdon (Baton Rouge: Louisiana State University Press, 1992), 50–57.

32. Helen Regis, "Second Lines, Minstrelsy, and the Contested Landscapes of New Orleans Afro-Creole Festivals," *Cultural Anthropology* 14 (1999): 472–504.

33. Mark Schuller and Marilyn M. Thomas-Houston, "Introduction: No Place like Home, No Time like the Present," in *Homing Devices: The Poor as Targets of Public Housing Policy and Practice,* ed. Marilyn M. Thomas Houston and Mark Schuller (Lanham, Md.: Lexington Books, 2006): 1–20.

34. Peter Sorant, Robert Whelan, and Alma Young, "City Profile: New Orleans," *Cities,* May 1984, 314–21.

35. Campanella, *Geographies of New Orleans,* 369–80.

36. Jennifer Ann Hay, "Restoring Cultural Capital through Preservation in the Holy Cross Historic District" (Ph.D. diss., Louisiana State University, 2014).

37. Roberto E. Barrios, "You Found Us Doing This, This Is Our Way: Criminalizing Second Lines, Super Sunday, and Habitus in Post-Katrina New Orleans," *Identities: Global Studies in Culture and Power* 17 (2010): 586–612.

38. Setha Low, "Maintaining Whiteness: The Fear of Others and Niceness," *Transforming Anthropology* 17 (2009): 72–92.

39. Mark Schuller, "Deconstructing the Disaster after the Disaster: Conceptualizing Disaster Capitalism," in *Capitalizing on Catastrophe: Neoliberal Strategies in Disaster Reconstruction,* ed. Nandini Gunewardena and Mark Schuller (Lanham, Md.: AltaMira Press, 2008), 27; Micaela di Leonardo, "Introduction: New Global and American Landscapes of Inequality," in *New Landscapes of Inequality: Neoliberalism and the Erosion of Democracy in America,* ed. Jane L. Collins, Micaela di Leonardo, and Brett Williams (Santa Fe, N.M.: School for Advanced Research Press, 2008), 5; Elizabeth Povinelli, "Shapes of Freedom: An Interview with Elizabeth A. Povinelli," *Alterites* 7 (2010): 90.

40. Ted Steinberg, "Can Capitalism Save the Planet? On the Origins of Green Liberalism," *Radical History Review* 107 (2010): 7–8.

41. Keith Marzalek, "City Announces First 17 Target Recovery Zones," *New Orleans Times Picayune,* March 29, 2007, blog.nola.com/updates/2007/03/city_announces_first _17_target.html (accessed November 12, 2014).

42. "1940: Charity Hospital in New Orleans Is Fully Occupied," *New Orleans Times Picayune,* November 13, 2011, www.nola.com/175years/index.ssf/2011/11/1940_charity _hospital_in_new_o.html (accessed November 20, 2014).

43. Henri Lefebvre, "The Right to the City," in *Writings on Cities,* ed. Eleanore Kofman and Elizabeth Lebastan (Oxford, U.K.: Blackwell, 1996), 101.

44. Gastón Gordillo, *Rubble: The Afterlife of Destruction* (Durham, N.C.: Duke University Press, 2014), 116.

45. Vincanne Adams, *Markets of Sorrow, Labors of Faith: New Orleans in the Wake of Katrina* (Durham, N.C.: Duke University Press, 2013).

46. Clara Irazábal and Jason Neville, "Neighborhoods in the Lead: Grassroots Planning for Social Transformation in Post-Katrina New Orleans?" *Planning, Practice & Research* 22 (2007): 131–53.

KATRINA IS COMING TO YOUR CITY

Storm- and Flood-Defense Infrastructures in Risk Society

KEVIN FOX GOTHAM

This essay examines the historical development of the New Orleans levee system and its catastrophic failure during the Hurricane Katrina disaster to advance understanding of the role of storm- and flood-defense infrastructures in the production of risk and vulnerability. As the deadliest and costliest failure of a major infrastructure project in U.S. history, the levee and floodwall breaches in the New Orleans region caused flooding in 85 percent of New Orleans, a disaster that contributed to over 1,400 deaths and estimated damages of approximately $150 billion. In New Orleans, Katrina flooded 12,000 business establishments (41 percent of the metropolitan area's total businesses) and 228,000 occupied housing units (45 percent of the metropolitan total). In the weeks after the storm, the Federal Emergency Management Administration (FEMA) distributed aid to over 700,000 households, including 1.5 million people directly affected by the storm. All told, 1.1 million people, 86 percent of the metropolitan population, lived in areas that were in some way affected by Katrina, through either flooding or other forms of damage.[1]

Two major concerns motivate my investigation. First, the failure of the New Orleans system has raised suspicions that other infrastructure projects and technologies may pose unique risks to cities and human health via exposure to environmental hazards and disasters. Books by William Freudenburg and colleagues (*Catastrophe in the Making*), Charles Perrow (*The Next Catastrophe*), John McQuaid and Mark Schleifstein (*Path of Destruction*), and Mike Tidwell (*The Ravaging Tide*) suggest that weakening and aging public infrastructure projects such as dams, bridges, roads, water supply systems, and transportation systems pose ominous threats to the safety and security of the nation.[2]

The American Society of Civil Engineers' (ASCE) "report cards" on the condition of infrastructure in the United States frame contemporary discussions of the crisis in quality and viability of public works. These annual reports assess sixteen types of infrastructure on the basis of their "capacity, condition, funding, future need, operation and maintenance, public safety and resilience." The ASCE's 2013 report gave D grades in eleven categories, including drinking water, hazardous waste, wastewater, roads, schools, transit, dams, inland waterways, levees, and energy. Except for solid-waste infrastructure, which received a B-grade, the other four types of infrastructure in the United States—rail, public parks and recreation, ports, and bridges—were in the C range.[3] Overall, deteriorating infrastructure and deferred maintenance have left a legacy of infrastructure that is of substandard quality, in numerous cases posing a danger of catastrophic failure.

Second, the failure of the New Orleans system offers an opportunity to examine the impact of specific organizations, government agencies, and policies in creating and distributing risks in the built environment. My intent is to examine the socially constructed nature of risk and risk assessment, and investigate the ways in which the very institutions assigned to regulate and reduce risk actually cause risks to expand through various technical procedures and organizational decision-making. Overall, I use the phrase "Katrina is coming to your city" as a metaphor to explain the increasing vulnerability of U.S. cities to disaster, to identify the connections between infrastructures and the production of risk, and to examine the impact of federal and state government policies in producing a wide range of hazardous, even deadly, consequences for U.S. cities. Katrina's devastation revealed the destructive potential not only of a hurricane but of decaying and poorly managed physical infrastructure. Indeed, the collapse of the New Orleans storm- and flood-defense system and recent reports of deteriorating dams and aging public water systems suggest a conception of public infrastructures as harbingers of insecurity that camouflage dangers for urban catastrophe. Because contemporary risks and vulnerabilities can be hidden from public view, the adverse effects of design defects or weak technology and infrastructure can lie dormant, ready to unleash a torrent of destruction the next time a storm, earthquake, or tornado strikes.

Storm- and Flood-Defense Infrastructures in Risk Society

Over the last two decades, the term "risk society" has entered the lexicon of major concepts used by scholars to explain the production of environmental and technological hazards and their linkages to postwar shifts in social organization and political economy.[4] Ulrich Beck argues that risks derive from hazards deliberately "induced by modernization itself" through formal organizations, government actions, and technologies designed to increase productivity and profit.[5] For Beck, the production of wealth and surplus accumulates hazards to the extent that unknown and unforeseen consequences of economic production "come to be a dominant force in history and society."[6] As he puts it, "new kinds of industrialized, decision-produced incalculabilities and threats are spreading within the globalization of high-risk industries. . . . [A]long with the growing capacity of technical options . . . grows the incalculability of their consequences."[7]

More recently, "social production of risk" approaches have emphasized how the consequences of particular definitions and meanings of risk arise from decision-making by formal organizations, political groups, and other powerful institutional actors. Risk implies uncertainty, probability, and the possibility of people being exposed to danger, harm, and other adverse outcomes in the present or future.[8] Central to this approach is that risks are "produced" through policies, sociolegal regulations, and routinized processes taking place within organizations, institutions, and government agencies. Social-production-of-risk approaches eschew a notion of risks as concrete, objective, and measurable, and examine the processes through which risks are socially defined and, as a result, which groups and interests benefit and suffer from prevailing definitions of risk. Such research recognizes the impact of political power and policy on social constructions of risk, the role played by economic constraints and pressures in political struggles over risk definition, and the effect of state actions and policy in the uneven sociospatial distribution of risks and vulnerabilities.

The insights of risk-society theorists and social-production-of-risk approaches reveal the ways in which urban infrastructures, including levee systems and flood-control projects, invite exposures to environmental hazards that pose catastrophic risks to human life. Infrastructures are complex systems of formal organizations, social networks, technologies, public policies, and financing that play an essential role in creating and regulating flows of capital, commodities, and people.[9] As products of sociolegal regulations and state policy, infrastructures aim to facilitate long-term investment, regulate market exchange, and anchor extralocal financial flows and networks of economic activity. As the skeletal support of communities and regions, infrastructures "bring heterogeneous places, people, buildings, and urban elements into dynamic articulation and interchange." As "congealed social interests," infrastructures and the policies and organizations that create them are never politically or socially neutral but express power relations and concerted efforts by particular groups and organized interests to control knowledge of risk and thereby extend their influence over urban space.[10]

In the United States, the U.S. Army Corps of Engineers has been the principal federal agency responsible for flood-defense infrastructure investment and construction. Early involvement of the corps in building these infrastructures came with the Flood Control Act of 1917, which authorized the agency to undertake flood-damage reduction activities across the nation. The Mississippi River Flood of 1927 spearheaded the passage of the Flood Control Act of 1928 and launched a massive federal program to strengthen flood control along the Mississippi, including the redesign and upgrading of levees.[11] The Flood Control Act of 1936 declared flood control a "proper" federal activity in the "national interest" and ushered in a modern era of federal flood-control investment. Over the decades, the corps has built 8,500 miles of levees and dikes, and constructed hundreds of smaller local flood-damage reduction projects.[12] Today, the corps maintains stewardship of 11.7 million acres of public lands, 14,500 miles of levees, 400 miles of shoreline protection, 702 dams, 926 harbors, and 13,000 miles of commercial inland waterways.[13] The corps' levees are located in 881 counties in the United States. These counties account for 37 percent of the nation's land area and contain more than 50 percent of the nation's population.[14] Although no one knows precisely how many people currently rely on levees for flood protection, 43 percent of the U.S. population lives in counties with levees, according to the ASCE.[15]

Organizational decision-making and political processes created conditions that led corps officials to downplay risks and thereby design the New Orleans hurricane and flood protection system—the Lake Pontchartrain and Vicinity Hurricane Protection Project (LP&VHPP)— using outdated engineering calculations and antiquated risk information. Funding conflicts with local sponsors (such as levee boards), combined with escalating project costs and constrained federal budgets from the 1970s forward, put pressure on corps officials to construct the hurricane and flood system according to original designs from the 1950s and early 1960s. Static funding translated into static designs as the New Orleans system became what one report called "a circa 1965 flood control museum" since the flood-control project was based on design assumptions and policy made in 1965.[16] In the ensuing decades, new knowledge of storm parameters, potential surge levels, climate change–driven sea-level rise, and coastal erosion was not incorporated into project design and construction. Once construction had begun, the corps focused on finishing the project as originally authorized and designed in 1965. Concerns for the affordability of the project and the urgency to get authorized protection in place thwarted changes in design, resulting in a series of decisions that over time spread flood and storm risk across a vast geographical area, thereby putting hundreds of thousands of people in harm's way.

Manufacturing Risk: The Origin of the LP&VHPP

The New Orleans hurricane- and flood-defense system has its origin in the 1950s when federal, state, and local governments began to develop plans to protect areas along the eastern and southern U.S. coasts from hurricanes and extensive flooding. In November 1962, the Corps of Engineers' New Orleans District completed an "Interim Survey Report" that outlined a comprehensive plan for preventing flooding in Greater New Orleans resulting from the "Standard Project Hurricane" (SPH), a hypothetical storm and performance standard the corps chose in 1950 as the recommended degree of protection. The SPH was developed by the corps "to provide generalized hurricane specifications that are consistent geographically and meteorologically for use in planning, evaluating, and establishing hurricane design criteria for hurricane protection works."[17] In conjunction with the U.S. Weather Bureau, the corps compiled data on all tropical storms of hurricane intensity within specific geographic zones over the period from 1900 to 1956. Using these data, the agencies created an index representing "the most severe combination of hurricane parameters that is reasonably characteristic of a specified geographical region, excluding extremely rare combinations."[18] For the coastal region of Louisiana, engineers expected an SPH to have a frequency of occurrence of once in about two hundred years, and represented "the most severe storm that is considered *reasonably characteristic* of the region" as of 1962.[19]

The original design for the city's hurricane-protection system was to control storm surge flowing into Lake Pontchartrain and bodies of water near downtown by building inlet barriers and canal floodgates. The corps planned to construct barriers to reduce SPH-driven surges along the lakefront. Levees and floodwalls in other locations would provide added protection to existing communities by containing hurricane surges. Following the devastation of New Orleans and other cities by Hurricane Betsy in September 1965, Congress passed the Flood Control Act of 1965, authorizing the Corps of Engineers to design and construct flood-defense structures, such as floodwalls and levees, to protect New Orleans and other cities. The LP&VHPP included lakefront barriers to reduce surge heights and prevent surges from entering the three outfall canals—the Orleans Avenue Canal, Seventeenth Street Canal, and London Avenue Canal—that penetrated into New Orleans from Lake Pontchartrain.[20] Cost-sharing rules for implementing a flood-control project required that nonfederal sponsors provide all lands, easements, and rights of way necessary for project construction, and these same entities were to assume long-term responsibility for the operation and maintenance of the completed project. As required by Congress, hurricane-project construction costs were to be shared with local sponsors according to a 70–30 match in which a local sponsor would pay for 30 percent of project construction costs.[21] The corps was responsible for project design and construction of the approximately 125 miles of levees. At the time of authorization in 1965, the corps estimated that the project would be completed by the mid- to late 1970s and cost approximately $80 million.[22]

Through the 1960s and 1970s, the corps' engineers and officials accumulated knowledge showing that the SPH model hurricane, used to design and evaluate the security and protectiveness of the LP&VHPP, was inadequate.[23] The hurricane storm surge, wave height, and catastrophic damage caused by Hurricane Betsy in 1965 and Hurricane Camille in 1969 raised doubts about the adequacy of the SPH. Hurricane Betsy caused extensive flooding throughout the New Orleans region, seriously damaging six thousand homes, grounding hundreds of barges in the Mississippi River, and inundating the Lower Ninth Ward with twelve feet of water.[24] Four years later, Hurricane Camille roared ashore a hundred miles east of New Orleans with 190-mile-per-hour winds and foot tidal surges between twenty-two and twenty-five feet.

Based on a review of previous calculations concerning the SPH, the New Orleans District Office reported that it was no longer the case that the LP&VHPP would contain SPH surges.[25] The corps' analyses after Hurricane Betsy found that existing levees and floodwalls would be insufficient in containing flooding even with the massive flood barriers as proposed by the LP&VHPP.[26] In 1979, the National Weather Service concluded that storms with parameters more severe than the SPH were more likely in the Gulf Coast region than previously thought possible.[27] A January 1979 report to the U.S. Water Resources Council raised doubts about the fiscal, economic, and environmental justification of the SPH as a design standard.[28] Despite the SPH's potential flaws and questionable assumptions, the corps went forward with designing the LP&VHPP and other flood-control and hurricane-protection projects around the nation using the model. Over time, the SPH came to represent not only a method for comparing storm risks between different geographic areas, but also a design standard that carried its own pledge of sufficient protection, security, and reliability.

Risk Controversies: Financial Noncommitment and Eroding Protection

Budget cuts, retrenchment, and chronic underfunding during the 1960s and later constrained planning and financing for the LP&VHPP and introduced a new conjuncture of dwindling resources and intense economic strain for the Corps of Engineers and local sponsors. The corps' national construction general budget in 1965 was over $6.5 billion. By 2005, forty years later, it had declined to about $3.6 billion, a dramatic fall in inflation-adjusted dollars. From the 1990s through the 2000s, federal investment in water-resources infrastructure was considerably lower than levels of the 1950s and 1960s. The agency's civil works budget fell from more than 1 percent of federal discretionary spending in the mid-1970s to between 0.6 and 0.5 percent in the years after 2000.[29] Annual appropriations for the corps' construction account fell from a $4 billion average in the mid-1960s (in 1999 dollars) to a $1.5 billion average for 1996 through 2005.[30] After the mid-1990s, competition for funds among corps projects was tight, and many projects that historically received appropriations were not included in recent budget requests. The data suggest that during this period rapidly increasing project costs were placing pressure on a stagnant federal water-development budget that had to be spread ever thinner for civil works projects nationwide.[31]

The multi-decade decline in funding for corps projects on a national level filtered down to the New Orleans District, resulting in irregular funding levels and decreasing financial support over time. The table below provides data on the annual administration budget request for the LP&VHPP from 1965 to 2006, the actual dollar amounts appropriated by Congress for the project in those years, and changes in funding in constant 2006 dollars. For over forty years, nominal appropriations for corps construction were very uncertain and unpredictable from year to year, with wild swings up and down. During the ten-year period before the Hurricane Katrina disaster, federal appropriations totaled about $128.6 million and corps reprogramming actions resulted in another $13 million being made available to the project. During this decade, appropriations generally declined from about $15–20 million annually in the earlier years to about $5–7 million in the last three fiscal years prior to 2006. While this may not be unusual, given the state of completion of the project, the appropriated amounts for these years were insufficient to fund new construction contracts.[32]

Escalating project costs combined with the inability of local sponsors to meet the burdensome 70/30 cost-share were a prominent source of conflict between local officials and the federal government over the LP&VHPP. Estimated costs skyrocketed to $200 million in 1971, more than three times the cost at authorization in 1965.[33] Efforts by elected officials to raise taxes to meet the local cost-share were overwhelmingly defeated by voters in 1970, 1972, and 1973.[34] After noting that the cost of the overall project had quadrupled since 1965, the Government Accountability Office (GAO) stated in 1976 that local sponsors would probably not be able to afford to pay their cost-shares.[35] In 1978, the president of the Orleans Levee District (OLD) expressed concerns about project financing and the inability to meet rising project costs. As the OLD president noted, "The Orleans Levee Board's share of the project approximates 67 percent of local participation. As of this date, if there are no further delays in the project, we estimate that we will have just enough money to pay our share.... Any delay will inflate the cost of the project in excess of the $400 million now estimated, and will place the cost beyond our ability to pay. What this probably means is that the project as it has been developed by the USCE [U.S. Army Corps of Engineers] by Congressional mandate is no longer viable."[36]

During these years, financially strapped local sponsors challenged corps designs and plans, arguing they were excessively costly, even to the extent of questioning whether the SPH surge was a real possibility. The corps' New Orleans District found itself defending the possibility of lakefront flooding and the SPH protection standard as it had

Construction General Funding for the LP&VHPP, 1979–2006

Year	Administration Budget Request (thousands of $)	U.S. House of Representatives Allowance (thousands of $)	U.S. Senate Allowance (thousands of $)	Final Appropriation (thousands of $)	Final Appropriation (thousands of constant 2006 dollars)
1966	NA	NA	$450	$538	$3,347
1967	$450	NA	$1,600	$1,600	$9,657
1968	$4,450	NA	$4,500	$4,086	$23,670
1969	$10,600	NA	$7,566	$6,269	$34,436
1970	$9,500	NA	$6,000	$8,500	$44,164
1971	$10,750	NA	$8,250	$14,500	$72,177
1972	$15,000	NA	$5,555	$7,755	$37,402
1973	$17,700	NA	$20,000	$20,000	$90,810
1974	$6,400	NA	$6,400	$6,400	$26,171
1975	$4,000	NA	$3,300	$3,300	$12,365
1979	$0	$0	$0	$0	$0
1980	$11,000	$11,000	$11,000	$11,000	$26,423
1981	$10,800	$10,800	$10,800	$10,800	$23,952
1982	$15,000	$15,000	$15,000	$15,000	$31,336
1983	$18,800	$18,800	$18,800	$18,800	$38,053
1984	$16,800	$16,800	$16,800	$16800	$32,597
1985	$17,500	$17,500	$17,500	$17,500	$32,788
1986	$25,000	$25,000	$25,000	$25,000	$45,985
1987	$16,000	$16,000	$16,000	$16,000	$28,394
1988	$17,000	$17,000	$17,000	$17,000	$28,970
1989	$40,000	$39,898	$39,898	$39,898	$65,683
1990	$39,898	$39,898	$39,898	$39,898	$61,541
1991	$11,655	$11,655	$11,655	$11,655	$17,251
1992	$21,491	$21,491	$21,491	$21,491	$30,491
1993	$11,607	NA	NA	$19,307	$27,061
1994	$9,619	$24,119	$24,119	$24,119	$32,809
1995	$10,000	$12,500	$12,500	$12,500	$16,535
1996	$7,848	$11,848	$11,848	$13,348	$17,150
1997	$4,025	$18,525	$18,525	$18,525	$23,268
1998	$6,448	$22,920	$16,448	$22,920	$28,347

Construction General Funding for the LP&VHPP (*continued*)

Year	Administration Budget Request (thousands of $)	U.S. House of Representatives Allowance (thousands of $)	U.S. Senate Allowance (thousands of $)	Final Appropriation (thousands of $)	Final Appropriation (thousands of constant 2006 dollars)
1999	$5,676	$18,000	$10,000	$16,000	$19,361
2000	$11,887	$16,000	$16,887	$16,887	$19,770
2001	$3,100	$8,100	$10,000	$10,000	$11,383
2002	$7,500	$13,500	$15,000	$14,250	$15,968
2003	$4,900	$9,000	$7,000	$7,000	$7,669
2004	$3,000	$5,000	$6,000	$5,500	$5,869
2005	$3,937	$7,500	NA	$5,719	$5,903
2006	$2,977	$2,977	$7,500	$4,000	$4,000

Sources: Budget figures from 1966 to 1975 come from Comptroller General of the United States, *Cost, Schedule, and Performance Problems of the Lake Pontchartrain and Vicinity, Louisiana, Hurricane Protection Project,* Report to Congress, PSAD-76-161 (Washington, D.C.: Comptroller General, August 31, 1976), 23, archive.gao.gov/f0402/098185.pdf (accessed February 27, 2011). All other years come from Woolley and Shabman, *Decision-Making Chronology for the Lake Pontchartrain & Vicinity Hurricane Protection Project, Final Report for the Headquarters, USACE* (2008), chap. 5, p. 4, library.water-resources.us /docs/hpdc/Final_HPDC_Apr3_2008.pdf (accessed December 15, 2014). Inflation-adjusted, constant-dollar amounts were calculated by the author.

been defined for the 1965 project authorization. State and local sponsors argued that the district's analysis had exaggerated the threat of hurricane-induced surges from the lake. As the GAO summed up in a 1982 report:

State and local sponsors generally believe the Corps has not pursued this project with the expediency necessary to protect the New Orleans area and that only another disaster resulting from a hurricane and heightened public interest would probably expedite project completion. The sponsors' major concerns were escalating project costs and their limited financial capability to pay for their share under either plan. Orleans Levee District officials believed that the Corps' standards may be too high for what is really needed for adequate protection and for what is affordable by local sponsors.... They recommended that the Corps lower its design standards to provide more realistic hurricane protection to withstand a hurricane whose intensity might occur once every 100 years rather than building a project to withstand a once in 200- to 300-year occurrence. This, they believe would make the project more affordable, provide adequate protection, and speed project completion.[37]

Public opposition had been building from the inception of the LP&VHPP, and by the mid-1970s coalitions of wetlands preservationists, environmental organizations, and elected officials were calling for an end to the plan. A 1975 public meeting conducted at the University of New Orleans drew dozens of protestors, many of whom denounced the corps' levee-building plans for their potential adverse effects on the lake environment. A spokesperson from the League of Women Voters proclaimed, "inasmuch as this project has been rejected by the voters on three separate occasions, we feel it is an affront to the public to proceed with it."[38] In 1976, a coalition of wetlands preservationists, environmental organizations, and local fishermen sued the corps, alleging that the final environmental impact statement (EIS) was not compliant with the requirements of the National Environmental Policy Act of 1969. The group leading the effort, Save Our Wetlands, modified the suit on March 8, 1976, to include allegations about the inadequacy of economic analysis and the inability of the Orleans Levee District to provide local assurances and meet the cost match.[39] On December 30, 1977, the U.S. District Court agreed with the claims of the plaintiffs and issued an injunction that prevented the corps from conducting any further work on the LP&VHPP until it had prepared an adequate EIS.

By this time, state and congressional officials had come out in opposition to corps plans, challenging claims that the LP&VHPP was necessary to protect New Orleans. Congressman Robert Livingston (R-La.) celebrated the injunction by saying, "we are losing an expensive bad plan." Echoing dominant criticisms from environmentalists, Livingston pointed out that the corps' plan to build barriers "presented an economic and environmental hazard to New Orleans, posing the possibility that the lake could turn into a dead sea."[40] Governor Edwin Edwards suggested the "abandonment of the barriers as unnecessary and as possibly dangerous to the environment." This oppositional stance marked an about-face for the government, which had previously urged environmentalists to drop their objections against the barriers.[41] As a *Times-Picayune* editorial remarked in July 1977, "The barrier concept does not have the support of the public and never did. The Orleans Levee Board tricked the public—urging them to vote for a three-mill tax while assuring them no portion of the tax would be used on the barrier structures. Shortly after the tax was approved, it was learned that the Levee Board ... had found a way to use the money for the barriers. ... There is no proof that the barriers will work to reduce hurricane damage or that alternatives are not better. ... Until studies are done to prove that barriers will have no adverse environmental effect on the ecology of the lake, no construction should take place. The Corps should cease and desist—now."[42]

Funding and policy constraints affected the corps' decisions. Raising the levees or armoring them against a storm bigger than the original SPH would have required a new authorization from Congress, or at least a post-authorization change approval. In addition, any proposed changes to the project design would need agreement by local sponsors to meet the requisite cost-sharing.[43] The logic for this longstanding approach was that all the segments of the entire LP&VHPP needed to be compatible and complete before requests for new authorization and funding could be made to Congress.

Seven years after the 1977 injunction, the corps officially abandoned the priority given to building barriers to control water flow into Lake Pontchartrain. In 1984, the corps announced a new plan, the High Level Plan, consisting of higher lakefront levees instead of barriers to control storm surges. The corps did not reevaluate or modify any other elements of the original LP&VHPP plan. By the 1990s, the New Orleans District recognized that accumulated new knowledge— including that related to subsidence, sea level rise, revised SPH central pressure, and more severe hurricanes as well as advances in computer surge-modeling and understanding—meant that the project that could be constructed within the estimated cost might not provide the authorized degree of protection.[44] According to a 1994 New Orleans District request to reevaluate the degree of protection that the LP&VHPP was providing to the region:

> The project was formulated in the 1950s and 1960s using the technology available at that time. Numerical models and high speed computers now allow for the more complete analysis of the physics of hurricane storm surge and wave action. Many reaches of the project are nearing completion. . . . [P]henomena such as sea level rise, deltaic subsidence and datum changes give cause for reanalysis. Our benchmark policy . . . when applied to project construction exacerbates the problems associated with sea level rise, deltaic subsidence and datum changes. These factors were not accounted for in the project formulation and have detrimentally affected the project's performance. . . . The trend in subsidence will continue. The degree of protection afforded by the project appears to be deficient in some reaches with the prospect of further deterioration over the remaining 100 year project life. In addition, topographic changes attributable to levee construction may have affected the hurricane stage relationship in Lake Pontchartrain.[45]

In sum, from the 1960s forward, corps officials recognized the potential implications of new storm information and later surge modeling and subsidence problems on the performance of the LP&VHPP. Despite this information and knowledge, the agency did not request authority and funding from Congress to budget to raise the floodwall heights or armor project structures. Despite recognition that the concept and application of the SPH were seriously flawed, no adjustments were made to the system before Hurricane Katrina roared ashore. Early warning signs of deficiencies and flaws persisted throughout development of the different components that comprised the LP&VHPP, but these signs were not adequately evaluated and acted upon.

Policy and Funding Constraints and the Normalization of Deviance

These budgetary and policy decisions by elected leaders and corps officials affected the development of the LP&VHPP, illustrating the "normalization of deviance" in which officials used obsolete and antiquated knowledge about risk to build a hurricane- and flood-control system. In her oft-cited book, *The Challenger Launch Decision,* Diane Vaughan examines the organizational and cultural processes under which NASA officials and managers made decisions to accept low-level risks and incorporated the decisions into shuttle design.[46] Competition for scarce financial resources contributed to "production pressures" that in turn led to the risky launch decision. Scarcity of resources can set the stage for accidents and disaster when officials respond with cost/safety tradeoffs. Budget retrenchment and cost-cutting, in particular, can increase risk, intensify danger, and compromise safety. In the case of the corps, the larger political and budgetary situation constrained decision-making and limited the actions that could be taken to address problems. Policy and funding constraints affected the corps' organizational culture, which in turn shaped its definitions and meanings of risk. Decisions were made to control costs, and the corps required contractors to use outdated benchmark elevations during construction.[47] Budget cuts and federal disinvestment were the mechanisms that impeded the incorporation of new knowledge into project designs and thereby allowed designs to remain in a state of protracted stasis.

Moreover, as a politicized and vulnerable agency, the corps has historically been dependent on earmarked funds and key political players who at times have accepted the corps' ambitious proposals but then imposed strict budget limits. Funding for the civil works program has often been a contentious issue between the Executive Branch and Congress, with final appropriations typically providing more funding than requested regardless of which political party controls the White House and Congress. The corps generally maintains strong congressional support because of the direct water-resource benefits and indirect economic and political benefits of its projects. When instituted in the 1950s, the 30 percent nonfederal cost responsibility for hurricane-protection projects represented a significant local burden that was without precedent. As a result, the corps' New Orleans District, when planning and then implementing the LP&VHPP, became especially cognizant of the acceptability and affordability of any plan to local sponsors. Nonetheless, the district still had to be responsive to the expectations and requirements of the hierarchical organizational structure of the corps as well as Executive Branch leadership and the Congress. In short, chronic underfunding, combined with organizational routines and longstanding intergovernmental tensions, resulted in systemic problems in the corps that caused the institutional failures responsible for the Hurricane Katrina disaster.

The development and failure of the LP&VHPP casts a light on the linkages between storm- and flood-defense infrastructure planning and investment, power relations, and government agencies and policies in creating and distributing risks in the built environment. Organizational actors can create large-scale risks that, when actualized, affect entire communities and cities in disastrous ways, as we see in Roberto Barrios's essay on the Lower Ninth Ward in this volume. Like all organizations, the corps has multiple goals and finite resources, which means that key decisions about risk involve tradeoffs among different goals and strategies. The corps' performance and pursuit of goals do not take place in a vacuum but are influenced by the political environment in which the corps operates. This political environment includes funding decisions and resource actions, interaction with federal as well as local entities, and preferences of elected officials and organizational leaders whose directives govern corps activities. As noted by sociologist Kathleen Tierney, "Risks can be thought of as generated and managed through processes that involved decisions among competing goals, influences from broader organizational networks or fields of interaction, and intra-organizational interactions and signals."[48] In the case of the LP&VHPP, the corps' organizational culture, the political environment of funding and resource decisions, and organizational arrangements between the corps and the federal government and local sponsors became major drivers and intensifiers of risk and vulnerability to disaster.

Conclusion

Risk societies are characterized by a proliferation of hazards and risks that have roots in sociolegal regulations, government actions and policies, and organizational processes. Risks do not happen by accident, nor are they the result of nature or ecological or meteorological processes. Rather, risks are generated by public and private decisions involving the control and distribution of valued political and economic resources. Present-day disasters are actualizations of risk conditions that are the product of past decisions made by powerful actors and organized interests acting through various institutional ties and network structures. Thus, a key focus in analyses of the social production of risk is the impact of organizations, institutions, and infrastructures on the production of risk. As Tierney has recently pointed out, "the severity of risks and whether they increase or decrease over time are in large measure a function of the behavior of organizations and institutions most directly involved."[49] The key point is that risks increase and disasters occur not because individuals in organizations make mistakes or bad decisions but rather because the structures in which organizations and institutions operate become enabling environments for risky activities. All of the social-structural conditions and organizational and institutional processes responsible for the Hurricane Katrina levee breaches were known and foreseeable in advance of these failures. In this sense, these were "predictable failures."

Since the Hurricane Katrina disaster, the federal government has spent $15 billion in rebuilding and upgrading a shattered system that was known for decades by the corps to be inadequate in protecting people. On the plus side, the new Hurricane and Storm Damage Risk Reduction System (HSDRRS) provides deeper and more stable steel-sheet piling within floodwalls and contains massive surge barriers as well as closeable gates, along with bypass pumps and drainage pumps to remove internal rainfall. The new system offers Greater New Orleans a 100-year level of risk reduction, meaning reduced risk from a storm surge that has a 1 percent chance of occurring or being exceeded in any given year. Today, the flood-insurance standard is the risk category that engineers are using to design a system to protect people. Local corps officials contend that the current 100-year system is built to more demanding specifications than those laid out in 1965 for a 200- or 300-year storm. Critics contend that the corps has rebuilt a system to a lower standard of protection than its poorly built system that collapsed during Hurricane Katrina. "In effect, it's not really a standard that's been set for safety reasons, it's a standard that's been set for obtaining affordable flood insurance only," said Bob Turner, an engineer who is the regional director of the Southeast Louisiana Flood Protection Authority–East. Critics add that the improvements don't measure up to what science now knows would be the "most severe conditions" that can reasonably be expected in this region—storms worse than Katrina.[50]

Like the multi-decade building of the LP&VHPP, the future development of the HSDRRS will likely involve a contested, trial-and-error searching process to confront the problems of hurricane protection and flood control within an ecologically sensitive environment and political-jurisdictionally fragmented metropolitan region. Several issues are worth mentioning. First, past actions and decisions that accelerated project costs and contributed to completion delays for the LP&VHPP remain endemic to HSDRRS and in the way the federal government plans and manages hurricane and flood-control infrastructures. Second, stagnant federal funding for public infrastructure and an exploding backlog of authorized projects waiting for federal funding will most likely constrain HSDRRS project development and long-term maintenance.[51] Third, the separation of authorization and appropriations processes and the tendency by Congress to allocate funds to projects without regard to risk-reduction benefits or degree of completion will also affect the new system, though it is difficult to predict how and under what conditions. Fourth, generating the local funds to pay for the massive costs of long-term maintenance of the HSDRRS is a daunting challenge for local governments in the New Orleans region given its shrunken post-Katrina tax base.

Today, Gulf Coast communities are dealing with new water-management challenges posed by climate change and aging water systems with limited funding for maintenance and construction. Despite the ongoing challenges of operating and maintaining aging infrastructure under budgetary constraints, local governments and the federal government must address the challenge of managing climate-change risk in order to limit flood- and other disaster-risk exposure to communities. In addition, climate change—especially sea-level rise on the Gulf Coast—poses a significant financial risk to federal, state, and local governments, including but not limited to their role as owners, operators, and managers of extensive infrastructure vulnerable to climate impacts. The federal government annually invests billions of dollars in infrastructure projects that state and local governments prioritize and supervise through, for example, decisions on zoning and how to build vulnerable infrastructure such as levees and floodwalls. While formulating and implementing climate-change adaptation strategies to protect Gulf Coast water and flood infrastructure may be costly, governments and communities are increasingly recognizing that the cost of inaction could be greater. Increasing the nation's ability to respond to a changing climate can be viewed as an insurance policy against climate-change risks and a resilience-building strategy that will enhance the prosperity of communities.

NOTES

1. Brookings Institution, *New Orleans after the Storm: Lessons from the Past, a Plan for the Future*, Washington, D.C.: Brookings Institution, October 2005, www.brookings .edu/~/media/research/files/reports/2005/10/metropolitanpolicy/20051012_neworleans .pdf (accessed June 29, 2013).

2. William R. Freudenburg, Robert Gramling, Shirley Laska, and Kai Erikson, *Catastrophe in the Making: The Engineering of Katrina and the Disasters of Tomorrow* (Washington, D.C.: Island Press, 2009); Charles Perrow, *The Next Catastrophe* (Princeton, N.J.: Princeton University Press, 2007); John McQuaid and Mark Schleifstein, *Path of Destruction: The Devastation of New Orleans and the Coming Age of Superstorms* (New York: Little, Brown and Co., 2006); Mike Tidwell, *The Ravaging Tide: Strange Weather, Future Katrinas, and the Coming Death of America's Coastal Cities* (New York: Free Press, 2006).

3. American Society of Civil Engineers, Report Card for America's Infrastructure (2013), www.infrastructurereportcard.org/ (accessed December 1, 2014).

4. Ulrich Beck, *Risk Society: Towards a New Modernity* (London: Sage, 1992); Anthony Giddens, *The Consequences of Modernity* (Stanford, Calif.: Stanford University Press, 1990).

5. Beck, *Risk Society,* 22.

6. Ibid.

7. Ibid. Beck's early work did not consider natural catastrophes as major components of his risk society thesis. He regarded floods, earthquakes, and hurricanes, among other natural disasters, as "old risks," compared to new risks such as radioactivity and chemical pollution. Only in his *World Risk Society* (London: Polity, 1999) did he conceptualize natural hazards and catastrophes.

8. K. Hewitt, ed., *Interpretations of Calamity: From the Viewpoint of Human Ecology* (Boston: Allen & Unwin, 1983); D. S. Mileti, *Disasters by Design: A Reassessment of Natural Hazards in the United States* (Washington, D.C.: Joseph Henry, 1999); Kathleen Tierney, *The Social Roots of Risk: Producing Disasters, Promoting Resilience* (Stanford, Calif.: Stanford University Press, 2014); Kathleen Tierney, "Toward a Critical Sociology of Risk," *Sociological Forum*, 14, no. 2 (1999): 215–42; Piers Blaikie, Terry Cannon, Ian Davis, and Ben Wisner, *At Risk: Natural Hazards, People's Vulnerability and Disasters* (London: Routledge, 2004); William R. Freudenburg, Robert Gramling, Shirley Laska, and Kai T. Erikson, "Organizing Hazards, Engineering Disasters? Improving the Recognition of Political-Economic Factors in the Creation of Disasters," *Social Forces* 87, no. 2 (2008): 1015–38.

9. Stephen Graham and Simon Marvin, *Splintering Urbanism: Networked Infrastructures, Technological Mobilities and the Urban Condition* (London: Routledge, 2001); M. A. Pagano and D. Perry, "Financing Infrastructure in the 21st Century City," *Public Works Management & Policy* 13 (2008): 22–38; Abhijeet Deshmukh, Eun Ho Oh, and Makarand Hastak, "Impact of Flood Damaged Critical Infrastructure on Communities and Industries," *Built Environment Project and Asset Management* 1, no. 2 (2011): 156–75; William M. Leavitt and John J. Kiefer, "Infrastructure Interdependency and the Creation of a Normal Disaster: The Case of Hurricane Katrina and the City of New Orleans," *Public Works Management & Policy* 10 (2006): 306–14.

10. Graham and Marvin, *Splintering Urbanism*, 10.

11. John M. Barry, *Rising Tide: The Great Mississippi Flood of 1927 and How It Changed America* (1997; New York: Simon and Schuster, 2007).

12. Larry S. Buss, "Nonstructural Flood Damage Reduction within the US Army Corps of Engineers," *Journal of Contemporary Water Research & Education* 130, no. 1 (2005): 26–30.

13. Figures appear in U.S. Army Corps of Engineers, *USACE Overview*, LTG Thomas Bostick Commander Headquarters, USACE, March 26, 2013, www.usace.army.mil/Portals /2/docs/USACE_101_April_2013.pdf (accessed December 16, 2014).

14. Data from FEMA National Flood Insurance Program Files compiled by Ezra Pound for Levees.org, levees.org/ (accessed January 5, 2010); Ezra Pound, *Assessing the Benefits of Levees: An Economic Assessment of U.S. Counties with Levees* (December 23, 2009), commissioned by Levees.org, levees.org/wp-content/uploads/2010/01/USCoun tiesWithLeveesMainMap5_121009_559K.jpg; "Assessing the Benefits of Levees: An Economic Assessment of U.S. Counties with Levees," paper presented at "Challenges of Natural Resource Economics & Policy [CNREP]: The Third National Forum on Socioeconomic Research in Coastal Systems" (New Orleans, May 26–28, 2010).

15. The American Society of Civil Engineers estimates that there may be as many as 100,000 miles of levees. More than 85 percent are thought to be locally owned, and the rest are overseen by the Corps of Engineers or other federal or state agencies. FEMA estimates that levees are located in roughly 22 percent of the nation's 3,147 counties. American Society of Civil Engineers, "So You Live Behind a Levee" (2010): 6, content.asce.org /files/pdf/SoYouLiveBehindLevee.pdf (accessed June 17, 2016).

16. Team Louisiana, *The Failure of the New Orleans Levee System during Hurricane Katrina,* report by Ivor Ll. van Heerden et al. for Secretary Johnny Bradberry, Louisiana Department of Transportation and Development, Baton Rouge, viii. See also Ivor Ll. van Heerden, "The Failure of the New Orleans Levee System following Hurricane Katrina and the Pathway Forward," *Public Administration Review* 67 (2007): 24–35.

17. National Hurricane Research Project, Howard E. Graham, and Dwight E. Nunn, *Meteorological Considerations Pertinent to Standard Project Hurricane, Atlantic and Gulf Coasts of the United States,* Report No. 33 (1959); J. J. Westerink and R. A. Leuttich, "The Creeping Storm," *Civil Engineering Magazine,* June 2003; Thomas O. McGarity and Douglas A. Kysar, "Did NEPA Drown New Orleans? The Levees, the Blame Game, and the Hazards of Hindsight," *Cornell Law Faculty Publications,* Paper 51 (2006), scholarship .law.cornell.edu/lsrp_papers/51 (accessed December 16, 2014).

18. See Harry S. Perdikis, "Hurricane Flood Protection in the United States," *Journal of the Waterways and Harbors Division, Proceedings of the American Society of Civil Engineers* 1 (February 1967).

19. Letter from the Secretary of the Army, July 6, 1965, "Re: Interim Survey Report on Hurricane Study of Lake Pontchartrain, Louisiana and Vicinity, United States Army Corps of Engineers," 67, library.water-resources.us/docs/hpdc/references.cfm (accessed October 25, 2016).

20. Hearings on Hurricane Protection Plan for Lake Pontchartrain and Vicinity before the Subcommittee on Water Resources of the House Committee on Public Works and Transportation, 95th Cong., 2nd Sess. (1978), p. 20 (testimony of Colonel Early J. Rush III). See also Government Accountability Office, *Hurricane Protection: Statutory and Regulatory Framework for Levee Maintenance and Emergency Response for the Lake Pontchartrain Project* (Washington, D.C.. Government Printing Office, December 15, 2005), p. 4.

21. Douglas Woolley and Leonard Shabman, *Decision-making Chronology for the Lake Pontchartrain & Vicinity Hurricane Protection Project: Final Report for the Headquarters, USACE* (2008), chap. 2, p. 19, library.water-resources.us/docs/hpdc/Final_HPDC_Apr3 _2008.pdf (accessed December 15, 2014).

22. Ibid., Executive Summary, 10.

23. The inadequacy of the SPH is detailed in a variety of reports. See, for example: U.S. District Court, Eastern District of Louisiana, "Re: Katrina Canal Breaches Civil Action Consolidated Litigation," No. 05-4182, *Order and Reasons,* Stanwood R. Duval Jr., U.S. District Court Judge; R. B. Seed et al., *Investigation of the Performance of the New Orleans Flood Protection Systems in Hurricane Katrina on August 29, 2005,* chap. 12 (section 12.5.1), www.ce.berkeley.edu/projects/neworleans/report/VOL_1.pdf (accessed Decem-

ber 16, 2014). In a historical analysis of the LP&VHPP, Woolley and Shabman reported, "The project degree of protection, as authorized, was protection against surges associated with the 1962-era SPH parameters for central pressure and wind speed. Project designs continued to be based on the 1962-era SPH parameters throughout the project history" (*Decision-making Chronology for the Lake Pontchartrain & Vicinity Hurricane Protection Project*, Executive Summary, 6–8).

24. U.S. Army Corps of Engineers, *Hurricane Betsy, September 8–11, 1965* (New Orleans: U.S. Army Engineer Office, 1965).

25. Woolley and Shabman, *Decision-making Chronology for the Lake Pontchartrain & Vicinity Hurricane Protection Project*, chap. 2, pp. 35–37.

26. Ibid., 37.

27. Ibid., 52.

28. U.S. Water Resources Council, *Report to U.S. Water Resources Council: Federal Agency Procedures for Project Design Flood Determination*, January 1979. References for the Decision-Making Chronology for the Lake Pontchartrain & Vicinity Hurricane Protection Project, library.water-resources.us/docs/hpdc/docs/19790100_Procedures_for _determining_project_design_flood.pdf (accessed December 11, 2014).

29. Office of Management and Budget, *Historic Tables, Budget of the United States, Fiscal Year 2006* (table 5.5), www.whitehouse.gov/omb/budget/fy2006/pdf/hist.pdf (accessed May 29, 2015).

30. Nicole T. Carter, *New Orleans Levees and Floodwalls: Hurricane Damage Protection*, CRS Report for Congress, Order Code RS22238, September 6, 2005, 5.

31. Nicole T. Carter, *Protecting New Orleans: From Hurricane Barriers to Floodwalls*, CRS Report for Congress, Order Code RL33188, December 13, 2005; Carter, *New Orleans Levees and Floodwalls*.

32. Carter, *New Orleans Levees and Floodwalls*.

33. Woolley and Shabman, *Decision-making Chronology for the Lake Pontchartrain & Vicinity Hurricane Protection Project*, chap. 2, p. 12.

34. Ibid., chap. 5, pp. 19–20.

35. Comptroller General of the United States, *Cost, Schedule, and Performance Problems of the Lake Pontchartrain and Vicinity, Louisiana, Hurricane Protection Project*, Report to Congress, PSAD-76-161 (Washington, D.C.: Comptroller General of the United States, August 31, 1976), archive.gao.gov/f0402/098185.pdf (accessed February 27, 2011).

36. Letter from Guy F. Lemieux, President, Board of Levee Commissioners of the Orleans Levee District, to the Honorable George Fischer, Secretary, Department of Transportation and Development, State of Louisiana, January 4, 1978, www.iwr.usace.army. mil/docs/hpdc/docs/19780104_OLD_cost_concerns_following_1978_injunction.pdf (accessed February 23, 2011).

37. General Accounting Office, *Improved Planning Needed by the Corps of Engineers to Resolve Environmental, Technical, and Financial Issues on the Lake Pontchartrain Hurricane Protection Project*, GAO/MASAD-82-39 (Washington, D.C.: Government Printing Office, August 17, 1982), 9.

38. Marietta Herr, League of Women Voters, *United States Army Corps of Engineers, Lake Pontchartrain, Louisiana, and Vicinity, Hurricane Protection Project,* Public Hearing, University Center Ballroom, University of New Orleans, February 22, 1975, 160, www.iwr .usace.army.mil/docs/hpdc/docs/19750222_Public_Hearing.pdf (accessed February 22, 2011).

39. *Save Our Wetlands v. Rush,* U.S. District Court, Civ. A. No. 75–3710 (E.D. Louisiana, December 20, 1977).

40. "Pontchartrain Hurricane Barrier Proposal Scrapped by Engineers," *Baton Rouge Morning Advocate,* June 8, 1978.

41. Ed Anderson and Larry Ciko, "Barrier Cut from Lake Storm Plan," *New Orleans Times-Picayune,* June 8, 1978.

42. "Hold Off on the Barriers," *New Orleans Times Picayune,* July 14, 1977, 22.

43. Woolley and Shabman, *Decision-making Chronology for the Lake Pontchartrain & Vicinity Hurricane Protection Project,* chap. 3, p. 28.

44. R. B. Seed et al., *Investigation of the Performance of the New Orleans Flood Protection Systems in Hurricane Katrina on August 29, 2005,* Independent Levee Investigation Team Final Report, July 31, 2006, www.ce.berkeley.edu/~new_orleans (accessed December 16, 2014).

45. Memorandum and Report from W. Eugene Tickner, Chief, Engineering District, to Commander, Lower Mississippi Valley Divisions, "Lake Pontchartrain, La. and Vicinity Hurricane Protection Project—Model Study, Request for Authorization," September 20 1994, 1–2, library.water-resources.us/docs/hpdc/docs/19940920_MVN_request_for _project_reevaluation.pdf (accessed December 11, 2014).

46. Diane Vaughan, *The Challenger Launch Decision: Risky Technology, Culture, and Deviance at NASA* (Chicago: University of Chicago Press, 1996).

47. Team Louisiana, *The Failure of the New Orleans Levee System,* ix.

48. Tierney, *The Social Roots of Risk,* 84.

49. Ibid., 44.

50. Bob Marshall, "New Orleans' Flood Protection System: Stronger Than Ever, Weaker Than It Was Supposed to Be," *The Lens,* May 15, 2014, thelensnola.org/2014/05 /15/new-orleans-flood-protection-system-stronger-than-ever-weaker-than-it-was -supposed-to-be (accessed May 29, 2015).

51. According to a 2012 National Research Council report on Corps of Engineers infrastructure, large portions of the corps' water resources infrastructure were built over fifty years ago and are experiencing various stages of decay and disrepair, making project maintenance and rehabilitation a high priority. The report also found that federal funding over the past twenty years has consistently been inadequate to maintain the corps' infrastructure at acceptable levels of performance and efficiency (National Research Council et al., *Corps of Engineers Water Resources Infrastructure: Deterioration, Investment, or Divestment?* [Washington, D.C.: National Academies Press, 2012]).

AFTERWORD

TED STEINBERG

The sea is knocking at the doors of coastal cities all across the world. From the bustle of New York to the canals of Amsterdam, from the shopping malls of Guangzhou to the slums of Mumbai, from Kolkata to Kuala Lumpur, London to Lima, the water is rising and perennial inundations are running up quite a bill, estimated to be roughly $6 billion a year in the 136 largest coastal cities. That coastal floods are a problem at a time of global warming and rising sea levels may not be surprising, but who would have thought that a fifth of those losses could be accounted for by just two metropolitan areas: Miami and New Orleans.[1] It is hard to escape the conclusion that there is something exceptional—exceptionally dangerous, that is—about the South's approach to the sea. Why do the Magic City and the Crescent City stand out in this way? How did the prospect of flood disaster come to loom so large over two of the most important cities in the region? What propelled them into the front rank of the planet's most vulnerable urban agglomerations?

According to the Insurance Information Institute, Hurricane Katrina is the costliest catastrophe in recent U.S. history with over $48 billion in losses (2014 dollars). The disaster claimed so many lives that the state of Louisiana never finished counting the dead. Florida, meanwhile, ranked number one in the United States in terms of insured catastrophe losses during the period from 1985 to 2014. Hurricane Andrew in 1992, the third costliest catastrophe (after Katrina and the September 11 attacks), and Hurricane Wilma in 2005, the seventh costliest, delivered their harshest blows in Miami-Dade County.[2]

Why Miami and New Orleans should share such a dark history might at first glance seem to be the product of environment alone. Both are located on former wetlands in marginal environments for building. About half of New Orleans now rests below sea level, as much as twelve feet below in some areas. South Florida, for its part, is flat as a pancake, with the average elevation in Miami-Dade County only six feet above the sea. Geology, too, is not working in favor of longevity. Porous limestone, shot through with holes filled with water, rests beneath the region.[3] But the similarities go far beyond topography and geologic facts. Both cities and the surrounding regions suffered through disasters in the 1920s that proved to be turning points. In the aftermath of those calamities, both metropolitan regions took the control of nature to new heights. It is no accident that two of the most flood-prone places on the entire planet have experienced some of the most massive efforts ever dreamed of to dominate the natural world.

* * *

When the French arrived in the Mississippi Delta in the early eighteenth century to stake claim to a crescent of land along a levee, the region had passed through a seven-thousand-year growth spurt. Before the arrival of the French, the Mississippi River had tacked back and forth, shifting course and, in the process, depositing layers of sediment and building up the land that the French had come to conquer. A cypress swamp sprawled out north of the new town, and beyond it existed Lake Pontchartrain. Water was literally everywhere the French settlers looked.

What attracted the French to the site was the location near the mouth of a major river combined with access to the sea. They arrived of course in a world wedded to maritime commerce and applied much the same logic that the Dutch employed roughly a hundred years earlier in locating New Amsterdam at the mouth of the Hudson River. They no doubt had no idea what a perilous enterprise they had helped to launch.

Dealing with unruly rivers was nothing new to the Europeans. To handle the Danube and the Volga, Europeans had built artificial levees; the French at New Orleans did the same. Indeed, levee building evolved into a monomania along the Mississippi by the middle of the nineteenth century, when the U.S. government stepped in and passed legislation that ceded swampland to states so that they could sell it off to finance more levee building. But confining the river required higher and higher levees and, with no organization overseeing the enterprise, the result was a jumble of building exploits that left the region vulnerable. So in 1879 Congress assigned the task of supervising the river to the Mississippi River Commission. Although the commission included some civilians, it was the U.S. Army Corps of Engineers that dominated the decision-making. Controlling the river with levees remained paramount to the corps' approach.[4]

By the early twentieth century, as New Orleans was passing through the Progressive Era of reform, city leaders began an assault on the cypress swamp, concocting an entirely new drainage system of canals and channels serviced with pumps to keep the city nice and dry and allowing real estate development to proceed apace. Taxable property almost doubled in value between 1890 and the outbreak of World War I. The year following—1915—a category four hurricane barreled through the region. The storm surge from Lake Pontchartrain killed 275 people. Nevertheless, the storm did nothing to unnerve New Orleans's boosters who, as Andrew Horowitz has argued, took the position that, if anything, the storm proved that since the city was—though waterlogged— still standing, the best course of action was for real estate and builders to continue on their merry way.[5]

Nevertheless, the challenges of building in a marginal environment remained. In 1927, a deluge displaced nearly a million people in the Mississippi River valley. To save New Orleans, the Corps of Engineers repaired to St. Bernard Parish, downstream from the city, determined to blow a hole in the levee and thereby, they thought, save the Crescent City. The massive explosions utterly destroyed the eastern portion of the parish, home to many Isleño fur trappers. Out of the disaster there emerged a change in the levees-only policy. The new approach was a colossal exercise in controlling nature that came to be called the Mississippi River and Tributaries Project. It would supplement levees with a set of new weapons for gaining control over the river, including bypass floodways, storage reservoirs, dikes, channel improvements— indeed, an entire armamentarium for shortening the river's course to the sea and keeping it locked in a straitjacket in which it could not imperil the port of New Orleans and development more generally in the delta. When a great flood struck in 1937, the authorities opened the new spillway at Bonnet Carré, upstream from New Orleans, and the city escaped unscathed.[6]

Ten years later, in 1947, a hurricane produced a storm surge that came over the top of the levees on Lake Pontchartrain. The disaster prompted the state of Louisiana to prevail on the federal government to bring its massive power and money to bear on the New Orleans flood problem. Studies began in 1955, but it took the destruction carried out by Hurricane Betsy ten years later—which caused floodwaters to race through the city, especially the Lower Ninth Ward, which had gone through the process of white flight—to compel the government to start building the Lake Pontchartrain and Vicinity Hurricane Protection Project. Tragically, the project did not come anywhere close to living up to its name. The project was ill-conceived, did not offer the advertised level of protection, did not conform to standard engineering practice, and was maintained, one critic has written acidly, "like a circa-1965 flood control museum."[7]

In this sense, Hurricane Katrina was simply a disaster waiting to happen. Massive intervention by the state, which essentially shared the risk of life in the delta with taxpayers throughout the country, has created a precarious city that cannot be saved by pouring more concrete alone. The reason is simple. Over the last eight decades or so, the seven-thousand-year-old process of land building has been turned on its head; the wetlands protecting the city have been decimated. All the dams and other water-control structures built throughout the Mississippi River valley have cut the sediment load in the river since 1950 by more than half if not more. Meanwhile, the drilling for oil and gas, a mainstay of the Louisiana economy, has permitted saltwater to intrude into the freshwater wetlands, destroying them.[8]

The wetlands devastation has been aggravated by boondoggles such as the Mississippi River Gulf Outlet Canal, designed to shorten the distance between the Port of New Orleans and the Gulf of Mexico. Completed in the mid-1960s, the canal not only destroyed immense amounts of wetlands as a result of both construction and saltwater intrusion, but also worked to funnel floodwater into the city during hurricanes Betsy

and Katrina. Construction of the canal alone is estimated to have an-
nihilated wetlands equivalent to three and a half times the size of the
entire island of Manhattan. The canal is now closed. But the loss of wetlands along the coast of Louisiana continues at a rate of roughly one
football field per hour, leaving New Orleans bereft of a frontline defense
against the sea.[9]

* * *

In 1880, at about the time the federal government stepped in to oversee
the so-called improvement of the Mississippi, New Orleans had a population of 216,000. The population of the entire state of Florida was only
about 50,000 larger. South Florida was slow to be developed because
too much water stood in the way. Surrounded on three sides by the sea,
the southern part of the state receives over fifty inches of precipitation
on average per year (1986–2015). Drainage is a problem aggravated by
the underlying geology, which has created a high water table.[10]

Approximately five thousand years ago, rising sea level and more
moisture created a world of plants and animals favorable to human occupation. But in the sixteenth century, European contact precipitated
epidemic disease that proved devastating and led to the abandonment
of the land. Without human beings intervening in the management of
the land, the Everglades, as it came to be called, evolved into a water
world of lakes and rivers, marsh and swamp, crawling with alligators
and crocodiles and overrun with mosquitoes. In 1840, an American
soldier described the environment as "a vast sea filled with grass and
green trees."[11]

It was an improbable place for a city. Nevertheless, in the 1870s, a
Cleveland businesswoman named Julia Tuttle visited Biscayne Bay, an
estuary spanning more than thirty miles along the coast of southeastern Florida. An associate of John D. Rockefeller named Henry Flagler
also visited Florida in the 1870s. Flagler would venture into the hotel
business; to bring visitors to his newly created paradise, he also invested in the Florida East Coast Railway. Together, Tuttle and Flagler
collaborated on a plan to build a railroad south to Biscayne Bay. In the

spring of 1896, they accomplished their goal, and later that year the city of Miami was incorporated. The city was named for the Mayaimi Indians, who lived around the state's largest lake—now named Lake Okeechobee—located about ninety miles to the north.[12]

Miami Beach was once just a sliver of sand bordered on the west by Biscayne Bay. A mangrove swamp unfolded across this reach. That is, until a flamboyant automobile magnate named Carl Fisher entered the picture. In 1913, Fisher and his associates had the mangrove swamp dredged. They then built a bulkhead and pumped in sand from the bottom of the bay to create a new, more substantial landmass that they divided into lots and sold off as real estate. Miami Beach was incorporated in 1915; a decade later, more than fifty hotels and almost two hundred apartment houses sprung up on what had once been a very tenuous barrier island.[13]

Urbanization's first major test came in 1926 when a category-four hurricane whipped through South Florida. At least a hundred people died in Miami, and fifteen thousand were left homeless. But much as had happened in New Orleans after the 1915 disaster, South Florida's boosters moved to minimize the disaster. The *Miami Herald* predicted that before too long "we will have ceased to think or talk of the hurricane." Instead, the city "will be looking forward, not backward." That is, to more growth and development.[14]

To the north of Miami, the wheels of progress also turned. During the Progressive Era, Napoleon Bonaparte Broward staked his reputation on the draining of the Everglades, overseeing the construction of several canals and, more important, convincing Floridians that there was no more time to second-guess the wisdom of tackling this obstacle in the way of the state's future. Control of nature it must be. Some four hundred miles of canals and levees had been built by the end of the 1920s. And valuable farmland, worked by black laborers, began to take command of the landscape around Lake Okeechobee.

Some preservationists took issue with the plan. Charles Torrey Simpson, for example, wondered whether "the world is any better off because we have destroyed the wilds and filled the land with countless human beings."[15]

As the Mississippi Valley was recovering from its most devastating flood, South Florida had its own firsthand experience with the perils of life on the edge of the sea. The 1926 hurricane had caused considerable death and destruction in Moore Haven, a town located on the southwestern shore of Lake Okeechobee, as the dike holding back the lake's waters gave way. Two years later, as many as twenty-five hundred people died—mainly black migrant workers—when the jerry-rigged muck wall restraining the waters of Lake Okeechobee again yielded, drowning its victims. Conquering nature in the Everglades proved no easy task, one that perhaps only the power and money of the federal government could address. At least that was the conclusion of Congress, which authorized the Corps of Engineers in 1930 to build a levee along Lake Okeechobee to a height of thirty-one feet above sea level. The structure, christened the Herbert Hoover Dike, was completed in 1937. The dike paved the way for the rise of modern agribusiness in the Everglades. Under the leadership of Charles Stewart Mott, the U.S. Sugar Corporation would capitalize on the dike and the rich Everglades muck to create an empire in this former subtropical wetland.[16]

The September 1947 hurricane that hammered New Orleans and prompted federal intervention in the city's hurricane protection also caused significant harm in Florida. Though the Hoover Dike held tight, the storm combined with another hurricane later in the fall to produce the worst flooding in a generation. And just as had happened in New Orleans, the flooding sparked federal involvement. In 1948, Congress authorized the Central and Southern Florida Project. For the next twenty years, the Corps of Engineers set to work engineering "the most elaborate water control system ever built, the largest earthmoving effort since the Panama Canal." The corps erected a massive levee down the eastern side of the Everglades to protect land closer to the coast from the perils of floods coming off the marshland. It also carved out hundreds of thousands of acres south of Lake Okeechobee and managed the land in the interests of agriculture. Sugarcane evolved into the dominant crop. The U.S. Geological Survey explained that the "project makes it possible for over five million people to now live and work in the 18,000 square mile area which extends from south of Orlando to Florida Bay."[17]

Again, there was resistance from some conservationists. "Quit being so land-hungry that Nature is left no place to store rainfall," said Ernest Lyons, an outdoor writer. Lyons indicted the idea of "calling on Government to be the very God, by the creation of a huge artificial system of dams, pumps, man-made lakes, and controls which must be maintained in perpetuity."[18]

The goliath scheme amounted to a monumental redistribution of acute local risk to taxpayers across the country—and, of course, a monstrous subsidy to Florida real estate and agribusiness. Property values in the district now under government management ballooned from $1.2 billion in 1950 to nearly $16 billion twenty years later.[19]

The project had stunning ecological impact. In the 1960s, after the Cuban Revolution prompted an embargo on sugar, the Corps of Engineers played god with the Kissimmee River. The watercourse once meandered more than a hundred miles—with water sloshing into the floodplain, which at points spanned two full miles wide—before flowing into Lake Okeechobee from the north. To dry out the land, the corps radically shortened the river by about half, plunging a thirty-foot-deep canal through the heart of the valley and destroying cnough wetlands to carpet Manhattan twice over.[20] Herons, wood storks, and largemouth bass all felt the effects. The sugar barons benefited immensely.

The Everglades are now half the size they once were. To make way for the six million people who live in the Miami metropolitan area, for Disney World, and for Big Sugar, the crocodiles, mink, fox squirrels, egrets, herons, and snail kites had to move over. Tens of millions of tourists flock to the region every year. Growth has continued apace, though the place has suffered through significant natural disasters, including Hurricane Andrew in 1992, a storm so fierce it destroyed 97 percent of the more than ten thousand mobile homes located in Miami-Dade County. Were a storm akin to the 1926 hurricane to visit the region again, the current estimate is that it would result in $125 billion in insured losses, or more than two and a half times the losses from Hurricane Katrina.[21]

* * *

The driving force behind the Gulf South's vulnerability to coastal flooding is the capitalist state. As historian Donald Worster has explained, the modern capitalist state is made up of both a private and a public sector. In Worster's model, which he applied to the history of water in the arid American West, the private sector is comprised of agriculturalists. And that remains true in our story here as well. But agribusiness is too narrow a definition of the private sector if we want to understand how the control of water paved the way for urban real estate and building in sprawling metropolitan areas. As for the public sector, it remains the same as in Worster's model: the technocrats bent on dominating nature and the representatives who empower them to carry out the task. The control of nature under this arrangement has one overriding goal, and that is capital accumulation.[22]

Nature, as Worster noted, has no intrinsic meaning in such a world. It is instead subjected to instrumental reason, that is, control for the sake of more growth and capital accumulation.[23] But push this form of reason too far and the contradictions of capitalist growth begin to emerge as they now have in the concentration of global flood risk at just two small points—New Orleans and Miami—on the planet.

The big question in Florida is this: Can enough canals to span all the way from New York to Las Vegas, more than sixty pump stations siphoning away, as well as two thousand structures designed to control the flow of water, defend South Florida, which is surging demographically, from the projected rise in sea level? For New Orleans the question is similar: With wetlands, the first line of protection from the sea, vanishing at a brisk rate, and places literally disappearing off the map, how much higher will the levees have to rise before people find that they are living out their lives at the bottom not of a bowl, as the city is often described, but of a pint glass?[24]

NOTES

1. Stephane Hallegate et al., "Future Flood Losses in Major Coastal Cities," *Nature Climate Change* 3 (September 2013): 802, 803; table 1 (p. 803) shows Miami and New Orleans ranked number two and four, respectively, in terms of average annual losses for

2005 with a combined loss (with protection) of nearly $1.18 billion, or 19.7 percent of the total of $6 billion in losses for all 136 major coastal cities.

2. Insurance Information Institute, "Top 10 Most Costly Catastrophes, United States," "Top Three States by Inflation-Adjusted Insured Catastrophe Losses, 1985–2014," www.iii.org/fact-statistic/catastrophes-us (accessed March 24, 2016); Carl Bialik, "We Still Don't Know How Many People Died Because of Katrina," August 26, 2015, fivethirtyeight.com/features/we-still-dont-know-how-many-people-died-because-of-katrina/ (accessed March 24, 2016).

3. Richard Campanella, "New Orleans Was Once above Sea Level, but Stormwater Drainage Has Caused It to Sink—with Deadly Consequences," January 28, 2016, www.nola.com/homegarden/index.ssf/2015/02/shifting_doorframes_cracking_d.html (accessed March 24, 2016); Elizabeth Kolbert, "The Siege of Miami," *New Yorker,* December 21 and 28, 2015, 45.

4. Ivor Ll. van Heerden, "The Failure of the New Orleans Levee System following Hurricane Katrina and the Pathway Forward," *Public Administration Review* 67 (December 2007): 25; John McPhee, *The Control of Nature* (New York: Farrar, Straus, Giroux, 1989), 36–37, 41.

5. Ari Kelman, "Boundary Issues: Clarifying New Orleans's Murky Edges," *Journal of American History* 94, no. 3 (December 2007): 701; J. D. Rogers, "Development of the New Orleans Flood Protection System prior to Hurricane Katrina," *Journal of Geotechnical and Geoenvironmental Engineering* 134, no. 5 (May 2008): 608; Andrew Deutsch Horowitz, "The End of Empire, Louisiana: Disaster and Recovery on the Gulf Coast, 1915–2002" (Ph.D. diss., Yale University, 2014), 4.

6. John M. Barry, *Rising Tide: The Great Mississippi Flood of 1927 and How It Changed America* (New York: Simon & Schuster, 1997), 208–9, 238–58, 423–26; Craig E. Colten, "Vulnerability and Place: Flat Land and Uneven Risk in New Orleans," *American Anthropologist* 108, no. 4 (December 2006): 732; McPhee, *The Control of Nature,* 43–44.

7. Rogers, "Development of the New Orleans Flood Protection System," 611; Colten, "Vulnerability and Place," 732–33; Heerden, "The Failure of the New Orleans Levee System," 28–30.

8. Nathaniel Rich, "The Most Ambitious Environmental Lawsuit Ever," *New York Times Magazine,* October 3, 2014, www.nytimes.com/interactive/2014/10/02/magazine/mag-oil-lawsuit.html (accessed March 24, 2016); Rogers, "Development of the New Orleans Flood Protection System," 603; Heerden, "The Failure of the New Orleans Levee System," 25–26.

9. Gary P. Shaffer et al., "The MRGO Navigation Project: A Massive Human-Induced Environmental, Economic, and Storm Disaster," *Journal of Coastal Research* 54 (Fall 2009): 207, 213. This article reports a loss of 21,000 hectares (51,892 acres) of wetlands. Manhattan is 14,610 acres. On current wetland loss, see Rich, "The Most Ambitious Environmental Lawsuit Ever."

10. South Florida Water Management District, "Rainfall Historical," www.sfwmd.gov/portal/page/portal/xweb%20weather/rainfall%20historical%20%28normal%20florida

%20annual%20rainfall%20map%29 (accessed March 24, 2016); Michael Grunwald, *The Swamp: The Everglades, Florida, and the Politics of Paradise* (New York: Simon & Schuster, 2006), 17.

11. David McCally, *The Everglades: An Environmental History* (Gainesville: University Press of Florida, 1999), 37, 59–62, 65 (quotation).

12. Grunwald, *The Swamp*, 105, 107.

13. Ted Steinberg, *Acts of God: The Unnatural History of Natural Disaster in America* (New York: Oxford University Press, 2000), 48–49, 50.

14. Ibid., 54, 55 (quotation).

15. Grunwald, *The Swamp*, 149, 175 (quotation); Steinberg, *Acts of God,* 59–60.

16. Eliot Kleinberg, "The Florida Flood That Accounted for the Most Deaths of Black People in a Single Day (Until Katrina)," *History News Network,* September 6, 2005, historynewsnetwork.org/article/15373 (accessed March 24, 2016); *An Act Authorizing the Construction, Repair, and Preservation of Certain Public Works on Rivers and Harbors, and for Other Purposes,* Public Law 520, *U.S. Statutes at Large* 46 (1930): 918, 925.

17. Grunwald, *The Swamp*, 218, 221; U.S. Geological Survey, "South Florida Restoration Science Forum," sofia.usgs.gov/sfrsf/entdisplays/restudy/ (accessed March 24, 2016).

18. Quotations are from Grunwald, *The Swamp*, 227, 228.

19. Ibid., 233–34.

20. U.S. Army Corps of Engineers, "Kissimmee River: Restoration Project," July 2012, www.saj.usace.army.mil/Portals/44/docs/Environmental/Kissimmee/Kissimmee_FS _June2015.pdf (accessed March 24, 2016); Grunwald, *The Swamp*, 268.

21. Steinberg, *Acts of God,* 92; Insurance Information Institute, "Estimated Insured Losses for the Top 10 Historical Hurricanes Based on Current Exposures," www.iii.org/fact -statistic/hurricanes (accessed March 24, 2016).

22. Donald Worster, *Rivers of Empire: Water, Aridity, and the Growth of the American West* (New York: Pantheon, 1985), 51.

23. Ibid., 52.

24. Kolbert, "The Siege of Miami," 45; Amy Wold, "Washed Away," *Baton Rouge Advocate,* April 29, 2013, theadvocate.com/home/5782941-125/washed-away (accessed March 24, 2016).

CONTRIBUTORS

ROBERTO E. BARRIOS is associate professor of anthropology at Southern Illinois University–Carbondale. During the past sixteen years, he has conducted ethnographic research on post-disaster reconstruction, with a specific focus on the neoliberal and modernist assumptions of disaster reconstruction policy and recovery planning, and the ways disaster-affected populations interpret, navigate, or contest these neoliberalisms and modernisms. His ethnographic case studies in Honduras, New Orleans, Mexico, and southern Illinois comprise the foundation of his forthcoming book *Governing Affect: Neoliberalism and Disaster Recovery.*

CHRISTOPHER M. CHURCH is assistant professor of history at the University of Nevada–Reno, where he teaches courses on French and Caribbean history, as well as classes on imperialism, disasters, and the digital humanities. He has published on race relations and citizenship in France and the United States, and his forthcoming book, *Paradise Destroyed: Catastrophe and Citizenship in the French Caribbean,* explores the social, economic, and political impact of disasters and civil unrest in the French West Indies.

URMI ENGINEER WILLOUGHBY is assistant professor of history at Murray State University. Since completing her doctorate at the University of California–Santa Cruz, she has held postdoctoral fellowships in comparative world history at Colby College and at the University of Pittsburgh's World History Center. She approaches histories of disease and medicine from a global and ecological perspective. Her research focuses on disease and ecology in the Mississippi Valley, Gulf South, and Caribbean, and draws connections between the southern United States and the colonial Atlantic and South Asia. Her book *Hurricane of the*

Human Frame: Yellow Fever, Race, and Ecology in Nineteenth-Century New Orleans examines the environmental, social, and cultural history of yellow fever epidemics in New Orleans in a global framework.

CINDY ERMUS is assistant professor of history at the University of Lethbridge in Alberta, Canada, where she teaches courses on early modern Europe, the Age of Revolutions, and the history of disasters. She has published on catastrophe and crisis management in eighteenth-century Europe and the Atlantic. Her current book project is a transnational study of the Plague of Provence of 1720, one of the last outbreaks of plague in Western Europe. By tracing a network of major eighteenth-century port cities, she explores the ways in which the crisis influenced society, politics, and commerce in neighboring regions and in the Atlantic and Pacific colonies.

ABRAHAM H. GIBSON teaches in the Department of History and Sociology of Science at the University of Pennsylvania. He is also a fellow in residence at the Consortium for History of Science, Technology, and Medicine in Philadelphia. His book *Feral Animals in the American South: An Evolutionary History* was published in 2016. He has also published extensively on a variety of topics related to southern history, environmental history, and the history of science. His forthcoming book examines the complex relationship between evolutionary biology and global diplomacy during the interwar period.

KEVIN FOX GOTHAM is professor of sociology and associate dean of grants, research, and graduate programs in the School of Liberal Arts at Tulane University. He has research interests in real estate and housing markets, post-disaster recovery and rebuilding, and the political economy of tourism. His most recent book, *Crisis Cities: Disaster and Redevelopment in New York City and New Orleans* (2014, coauthored by Miriam Greenberg), examines the federal response to the September 11 and Hurricane Katrina disasters. He is author of *Race, Real Estate, and Uneven Development* (2002; second edition, 2014), *Authentic New Orleans* (2007), *Critical Perspectives on Urban Redevelopment* (2001), and dozens of articles and book chapters on housing policy, racial segregation, urban redevelopment, and tourism.

ANDY HOROWITZ is assistant professor of history at Tulane University, where he specializes in modern American political, cultural, and environmental history. His dissertation, "The End of Empire, Louisiana: Disaster and Recovery on the Gulf Coast, 1915–2012," won the Southern Historical Association's 2015 C. Vann Woodward prize for best dissertation on southern history and Yale University's 2015 George Washington Egleston Prize for best dissertation on American history. He is currently writing a book on the history of the Katrina disaster.

GREG O'BRIEN is associate professor of history at the University of North Carolina–Greensboro and executive editor of the journal *Native South*. His books include *Choctaws in a Revolutionary Age, 1750–1830* (2002) and *Pre-Removal Choctaw History: Exploring New Paths* (2008). He has published articles on the history of the Native South in the *Journal of Southern History* and in essay collections and has written about engineer George Towers Dunbar Jr. and early American conservation in the essay collection *Backcasts: A Global History of Fly Fishing and Conservation* (2016). He is presently completing a book about the New Orleans Flood of 1849.

TED STEINBERG is Adeline Barry Davee Distinguished Professor of History and professor of law at Case Western Reserve University. A native of Brooklyn, New York, Steinberg has held fellowships from the Michigan Society of Fellows, the Guggenheim Foundation, the National Endowment for the Humanities, the American Council of Learned Societies, and Yale University. He has written six books in the field of environmental history and is the author, most recently, of *Gotham Unbound: The Ecological History of Greater New York* (2015).

INDEX

1926 hurricane, 87, 93–94, 96, 98, 189–90, 191

1928 hurricane. *See* Okeechobee Hurricane of 1928

A. aegypti mosquito. *See* mosquito

Accampo, Elinor, 1

Aden, Yemen, 50

Africa, 56, 61n73, 119

Agramonte, Aristides, 45

agriculture/agro-industry, 8, 81, 83–85, 87–90, 92, 94–97, 99, 100, 111, 113, 114, 134, 190–91, 192

Alabama, 1, 106, 113, 115

Albany, N.Y., 42

Albertini, A. Díaz, 46

Alexander, Jeffrey C., 79n58

alligators, 3, 15, 83, 188

American Red Cross, 92, 96

American Society of Civil Engineers (ASCE), 162, 181n15

American Society of Tropical Medicine, 55

Amsterdam, 184

Anthropocene, 4, 38

Antilles, 81, 84, 90, 97, 99

Arkansas River valley, 23

Army Medical Reserve Board, 45

Army Medical School, 45

Army Nurse Corps, 45

Asia, 2, 50, 116

Ayala, César, 81, 90

Bacillus icteroides, 45

Bacon, Francis, 107

bacteriology, 7, 39–41, 44, 45, 47, 54, 57n7

Bahamas, 89, 90, 93

Bahian Tropicalista School of Medicine, 56

Baltimore, Md., 45

banana industry, 98

Bangkok, Thailand, 50

Bankoff, Greg, 2

Barbados, 53

barrier (flood/water), 5, 8, 166–67, 172–73, 177. *See also* dike; floodgate; floodwall; levee; seawall

barrier islands, 86, 189

Batavia (Jakarta), 50

Bates, Marston, 107

bayous, 3, 137, 141; Bayou Bienvenue, 144, 150, 151–53, 155

Beard, Charles, 73

Beck, Ulrich, 163

beet sugar, 90, 98

Belize, 48, 53

Belle Glade, Fla., 94, 97

Belo, A. H., 76n6

Bemiss, S. M., 42

beriberi, 56

Bernard, Shane K., 112

biodiversity, 104, 106, 107

biogeographical realms, 106–7

biological invasions. *See* non-native species

Biscayne Bay, 188–89

Blagden, John, 70

Blakely, Edward, 132

Board of Public Works of the State of Louisiana, 18

Boca Raton, Fla., 94

boll weevils, 8, 114–16, 120

Bombay. *See* Mumbai

Bonnet Carré Spillway, 187

Boston, Mass., 42

Boyce, Rubert, 53

Brazil, 47–48, 51, 52, 53, 56

Brisbin, Lehr, 110

British Empire, 25, 46, 48, 50, 53, 54, 55, 56

Broward, Napoleon Bonaparte, 189

Broward County, 86

Brown, James (English immigrant), 70

Buffalo Creek flood, 74

Buhs, Joshua Blu, 114

Burmese pythons, 8, 116–19. *See also* snakes

Byron, Lord, 23

Calcutta. *See* Kolkata

Caribbean, 1, 6, 7, 10n2, 37–38, 39, 48–49, 53, 55, 80–83, 84, 85, 86, 88–92, 94, 97–99

Carr, Lowell J., 83

Carroll, James, 45, 46, 59n30

Carson, Rachel, 1, 113–14

Caruth, Cathy, 76

cattle, 68, 97, 113

census, 89, 131, 142

Central America, 37, 53, 81

Central and Southern Florida Project, 190

Chaillé, Stanford E., 42, 44

Charleston, S.C., 38, 42

Chase, Joshua C., 85, 87, 96

Cheesborough, Edmund, 72

chemical/toxic spill, 103, 180n7

Chile, 1

cholera, 34, 42

Citizens' Volunteer Ward Organization, 52

City Times (Galveston), 73

Civil War, 42

class inequality. *See* social inequality

Clewiston, Fla., 97

climate change, 1, 2, 3, 4, 9, 121, 165, 178–79, 184. *See also* sea-level rise

Coastwide Nutria Control Program, 112

Coates, Peter, 105

cocoa industry, 98

coffee/coffee cultivation, 26, 92, 96, 98

Colombo, Sri Lanka, 50

colonization/colonies, 39, 48, 49, 50, 53–56, 80, 88–89, 90–93, 94, 96, 106, 133, 138. *See also* British Empire; imperialism

Colorado State University, 152

Colten, Craig, 5

Columbian Exchange, 106

Columbus, Christopher, 82, 106

Commercial Bulletin (New Orleans), 20

cotton/cotton cultivation, 33, 71, 91, 115–16, 120

Coulter, John, 62–3, 65, 70, 71

Crescent City. *See* New Orleans

crevasse, 14, 16, 19, 20–21, 24, 27–31, 33–34

crocodiles, 118, 188, 191; Nile crocodile, 119–20

Crosby, Alfred, 121

Crossman, A. D., 18, 24, 30

Cruz, Oswaldo, 48

Cuba, 7, 44–47, 48, 51–55, 67, 84, 88, 89, 92, 97; Cuban embargo, 97, 191

Cuban Revolution, 191

cypress forests/swamps, 135, 137, 140, 141, 150, 151, 153, 185

Daily Crescent (New Orleans), 14, 20, 25

Daily Delta (New Orleans), 20, 22, 23, 24, 25

Danube River, 185

Davis, Mike, 96

Demerara, 53

Denmark, 97

Department of Military Hygiene at the U.S. Military Academy, 45

De Soto, Hernando, 110

Dickens, Charles, 23, 25

dike, 85, 88, 94, 96, 99, 164, 186, 190; Herbert Hoover Dike, 96. *See also* barrier (flood/water); floodgate; floodwall; levee; seawall

disaster: meaning/etymology of, 7, 10, 11n3, 15–16, 63–64, 66–67, 74–75, 83, 103–6, 119–22, 133–34, 177, 180n7. *See also specific disasters*

disaster studies, 4, 8, 9, 83, 103–4, 105, 106, 138

disease, 7, 34, 37–45, 48–51, 54–56, 91, 92, 103, 137, 188, 195; epidemics, 2, 37–39, 41–43, 45, 50, 52, 92, 103–4, 188. *See also specific diseases*

Disney World, 191

Dominican Republic, 88, 92, 97

Dorcas, Michael E., 116, 117

dracontiasis, 56

draining, of wetlands (including Everglades), 38, 47, 83, 84–86, 87, 89, 90, 94, 143, 188, 189; drainage-levee failure, 139; drainage system, 18, 19, 88, 121, 177, 186

drought, 1, 2

Dunbar, George Towers, Jr., 18–21, 24, 27, 30–31, 33, 34

Dynes, Russell, 83

earthquakes/seismic activity, 1, 2, 5, 8, 15, 83, 103, 104, 119, 163, 180n7

Edwards, Edwin, 172

Edwards, John Bel, 3

Egyptian Medical Congress, 50

Elton, Charles, 107, 108, 120, 121

England, 47. *See also* British Empire

Enterprise, Ala., 115

environmental injustice, 143. *See also* social inequality
environmental justice, 132
epidemics. *See* disease
Erikson, Kai, 74
erosion, 2, 3, 5, 8, 9, 10, 139, 165
ethnic inequality. *See* social inequality
Everglades, 83–85, 86–88, 94, 97, 99, 116–17, 119, 188, 189, 190–91; Everglades Drainage Commission, 85; Everglades Drainage District, 85
exotic-pet trade, 116, 118, 120

famine, 2
Fayling, Lloyd, 67–69, 71, 72
Federal Emergency Management Administration (FEMA), 161, 181n15; National Flood Insurance Program, 180n14
Finlay, Carlos, 44, 46, 47
Fiori, Januario, 48
fire, 1, 2, 103
fire ants, 8, 113–14, 120
First World War. *See* World War I
Fisher, Carl, 189
Flagler, Henry, 188
Flood Control Acts: of 1917, 164; of 1928, 164; of 1936, 164; of 1965, 166–67
floodgate, 166, 177. *See also* barrier (flood/water); dike; floodwall; levee; seawall
flooding/flood risk/flood control, 2, 3, 86, 93, 95, 96, 103, 137–41, 143, 144, 157, 161–67, 169, 175–79, 180n7, 184, 185, 186, 187, 190, 192; during Hurricane Betsy, 166, 167, 187. *See also* Hurricane Katrina; *names of specific floods*
floodplain, 131, 191
floodwall, 161, 166, 167, 174, 177, 179. *See also* barrier (flood/water); dike; floodgate; levee; seawall
Florida, 1, 3, 6, 8, 38, 51, 80–90, 92–97, 99, 106, 109, 110, 116–19, 137, 153, 184, 188–91, 192
Florida Bay, 190
Florida Department of Agriculture, 89
Florida Keys, 118. *See also* Key West
Florida State Board of Health, 95
Fort Pierce, Fla., 85
France, 47, 90, 98
French Guyane, 91
Fressoz, Jean-Baptiste, 5

Freudenburg, William, 161
Fritz, Charles, 64, 66–67, 74
Fukushima Daiichi Nuclear Power Plant, 2
fur industry, 112
fur trappers, 186

Galmot, Jean, 91, 101n38
Galveston, Tex., 38
Galveston City Commission, 72–73
Galveston flood. *See* Galveston hurricane
Galveston hurricane, 7, 62–76, 87. *See also* flooding/flood risk/flood control
Galveston Tribune, 65
gentrification, 145, 146, 155, 156, 157, 158
Germany, 47, 139
germ theory, 40
Gilmore, Glenda, 75
global warming. *See* climate change
Gordillo, Gastón, 155
Gorgas, W. C., 45, 46, 50, 51
Government Accountability Office (GAO), 169, 171
Great Depression, 81, 91, 98
Grenada, 53
Guadelupe (island), 80, 90–91, 92, 93, 98
Guam, 47
Guangzhou, China, 184
Guitéras, Juan, 44, 46
Gulf Coastal Plain, 109
Gulf of Mexico, 10n2, 44, 50, 62, 74, 80, 81, 90, 106, 109, 118, 138, 139, 140–41, 187

Haiti, 5, 97
Hall, A. Oakey, 41
Halstead, Murat, 66, 69
Hamburg School of Tropical Medicine, 48
Hardee, Thomas S., 42, 44
Havana, Cuba, 44–46, 47, 48, 50, 51
Havana Yellow-Fever Commission, 43
Hawaii, 47
hazard, 1, 2, 3, 4, 5, 8, 10, 11n3, 12n15, 16, 37, 38, 80–83, 84–85, 96, 104, 116, 120, 137, 139, 161–64, 172, 176, 180n7
Herbert Hoover Dike, 96, 190
High Level Plan, 173
Hispaniola, 82
historic preservation, 145, 147–48, 149, 156, 157
Hollander, Gail, 84
Holmes, Bernie, 147

Holmes, Jeanelle, 148
Honduras, 48, 53
Hong Kong, 50
Horowitz, Andy, 5, 186
horses, 68, 153
Howard, Leland O., 46
Howard, Philip A., 57n7
Hudson River, 185
hunting, 110, 111, 112, 117, 120, 153
Hurricane Andrew, 80, 184, 191
Hurricane and Storm Damage Risk Reduction
    System (HSDRRS), 177–78
Hurricane Betsy, 166, 167, 187
Hurricane Camille, 167
Hurricane Katrina, 1, 2, 4, 5, 7, 8–10, 80, 87, 131–
    34, 135, 137, 139, 141, 142, 144–47, 149–50,
    151, 152, 154, 155, 156–58, 161, 162, 169, 174,
    176, 177–78, 184, 187, 188, 191
hurricanes, 2, 3, 7, 11n3, 15, 17, 75, 82, 83, 87,
    91, 99, 103, 140, 162, 165–67, 171–75, 177–78,
    180n7, 186–87, 190; hurricane season, 80, 83.
    See also names of specific hurricanes
Hurricane Sandy, 1
Hurricane Wilma, 184

immigrants/immigration, 22, 38, 39, 46, 86, 89,
    96, 105, 121, 139. *See also* laborers
imperialism, 37, 47, 53, 54–56. *See also* British
    Empire
Industrial Canal (New Orleans), 131, 139–41
influenza, 92
infrastructure, 9, 84–85, 134, 161–64, 168, 176,
    177–79, 183n51
Ingold, Tim, 137
Inner Harbor Navigation Canal. *See* Industrial
    Canal (New Orleans)
Insurance Information Institute, 184
invasion biology, 104, 105–9, 119–21
invasive species. *See* non-native species
Ireland, 139
Irving, Washington, 23
Italy, 139

Jackson, Jeffrey H., 1
Jackson, John, 146
Jackson, Miss., 42
Jackson, Victoria, 131
Jacksonville, Fla., 84
Jamaica, 53

Japan, 2, 50
Jim Crow, 65
Johnstown Flood, 62, 65
Joint Committee on the Crevasse, 21
Jones, Walter C., 63, 72

Kalm, Peter, 107
Kendall, George Wilkins, 22
Ketchum, Ed, 68
Key West, Fla., 86, 89
King, Joshua R., 114
Kissimmee River, 191
Knowles, Scott, 5, 6
Koch, Robert, 40
Kolkata, 50, 184
Kuala Lumpur, 184
kudzu, 105

laborers, 8, 20, 85; agricultural wage labor, 81,
    88–89; African, 92; African American, 95,
    115, 142, 189, 190; Afro-Caribbean, 92, 99;
    industrial wage labor, 81; migrant workers, 8,
    88–99, 94, 96, 99, 190; unfree, 80; unpaid, 88
LaCapra, Dominic, 75
Lake Okeechobee, 8, 83, 85–86, 89, 92–97, 99,
    189, 190, 191
Lake Pontchartrain, 16, 17, 19, 20, 24, 131, 139,
    142, 154, 166, 173, 174, 185, 186, 187
Lake Pontchartrain and Vicinity Hurricane
    Protection Project (LP&VHPP), 165–76, 178,
    182n23, 187
Lange, Fabian, 114
Larson, Erik, 63, 76n1
Las Vegas, Nev., 192
Latin America, 56
Latour, Bruno, 136
Lazear, Jesse, 45, 46
League of Women Voters, 172
Lefebvre, Henri, 134–36, 137, 141, 142, 150, 151,
    152, 153
Leopold, Aldo, 107
leprosy, 56
Lester, Paul, 66, 71
levee, 9, 14–20, 24, 27, 29, 33, 96, 131, 133, 136,
    138–39, 141, 143, 144, 155, 161, 162, 164–67,
    169, 171–74, 177, 179, 181n15, 185–87, 189–90,
    192. *See also* barrier (flood/water); dike;
    floodgate; floodwall; seawall
Lewis, Ronald W., 154

Lima, Peru, 184
lionfish, 8, 118, 120
Lister, Joseph, 40
Liverpool School of Tropical Medicine, 47–48, 53
Livingston, Robert, 172
Lodge, David, 108
London, 184
London School of Tropical Medicine, 47, 54
looting/looters, 62–63, 64, 66, 68–71, 76, 78n32
Louisiana, 1, 3, 4, 10, 10n2, 18, 33, 40, 42, 51, 60n58, 106, 111–12, 115, 132, 133, 141, 149, 152, 166, 178, 184, 187, 188
Louisiana Recovery Authority (LRA), 132
Louisiana State Board of Health, 43, 51
Louisiana State University, 152
Low, Setha, 145
Low, Timothy, 121
Lower Ninth Ward. *See under* New Orleans
lynching, 71, 74
Lyons, Ernest, 191

Madagascar, 118
Make It Right Foundation, 154
Malaria, 33, 45, 46, 50, 54–55, 60n58, 61n74, 92
mangroves, 3, 86, 109, 189
Manhattan, 188, 191, 193n9
Manson, Patrick, 46, 50, 54–55
Marine Hospital Service. *See* U.S. Public Health and Marine Hospital Service
Marsh, George Perkins, 121
martial law, 63, 68, 69, 71, 72
Martin, John, 84, 87
Martinique, 91, 98
Marx, Karl, 38, 56n1
Masco, Joseph, 141
Mathewson, Kent, 10n2
Maxon, Harry, 65
Mayaimi Indians, 189
Mayer, John "Jack," 110
McClendon, Ward "Mack," 152, 154–55, 157, 158
McIlhenny, Edward Avery, 112
McKnight, Tom, 110
McNeill, J. R., 38, 56n1
McQuaid, John, 161
melaleuca tree, 83
Memphis, Tenn., 41, 42, 43, 70
Metairie Ridge, 17, 19, 24
Mexico, 37, 47, 48, 49, 51, 53, 71

Miami/Miami Dade County, 86, 87, 89, 93, 96, 184–85, 189, 191, 192, 192n1
Miami Beach. *See* Miami/Miami Dade County
Miami Herald, 189
miasma/miasmatic theory, 39, 40, 54
migrant workers. *See* laborers
military-medical complex, 53–55
militia/militiamen, 62, 63, 68, 69, 75
Minnesota, 70
Mississippi, 1, 4, 6, 106, 115; delta, 23, 114, 133, 137, 185; river, 14, 16–20, 22–24, 30–33, 42, 109, 131, 136, 138–40, 144, 154, 167, 185, 188; river basin, 10n2; river valley, 23, 37, 38, 41–42, 51, 186, 187, 190
Mississippi Free Trader and Natchez Gazette, 24
Mississippi River and Tributaries Project, 186
Mississippi River Commission, 186
Mississippi River Flood of 1927, 164, 186
Mississippi River Gulf Outlet Canal (MR-GO), 140–41, 187–88
Mississippi River/New Orleans Flood of 1849, 6, 14–25, 29, 31, 33–34
Mitchell, Timothy, 136
Mobile, Ala., 21, 38, 42, 45, 113
mobile homes, 191
Montserrat, 98
Moore Haven, Fla., 190
Moreira, Oscar Marques, 48
Moses, Wallace R., 85
mosquito, 3, 6, 7, 37, 39, 40, 43, 44–52, 56, 188; *A. aegypti* mosquito, 38, 46, 57n2, 59n33; *Anopheles* mosquito, 46, 54, 60n58, 61n74; *Culex* mosquito, 46; *Stegomyia* mosquito, 48–49, 52, 53, 59n33; anti-mosquito brigades/campaigns, 47, 48, 49, 50, 52
Mott, Charles Stewart, 190
Mouton, Alexandre, 18
Mumbai, 50, 184

Nagin, C. Ray, 132
National Aeronautics and Space Administration (NASA), 175
National Board of Health, 42, 43
National Environmental Policy Act (NEPA), 172
National Feral Swine Damage Management Program, 111
National Geographic, 65
National Research Council, 183n51
National Weather Service, 167

nativism, 105, 107, 108

natural hazard. *See* hazard

neoliberalism, 5, 137, 148–49, 155–57

New England, 67

New Jersey, 73

New Orleans, 6, 7, 9, 10, 10n2, 14, 16–19, 21–24, 32–34, 37–43, 50–54, 56, 131–32, 135, 137, 139–44, 146–49, 154, 157, 161–62, 165–67, 168–69, 171–73, 175, 177, 178, 184, 185–88, 189, 190, 192, 192n1; Almonaster Avenue, 139; Audubon (neighborhood), 132; Canal Street, 18; Carrollton (neighborhood), 17, 18, 19, 20; Central Business District, 18, 140; Charity Hospital, 149; Flood Street, 148; French Quarter, 140; Holy Cross (neighborhood), 144–47, 150; Jefferson Parish, 142; Lafayette (neighborhood), 17, 19, 20, 33; London Avenue Canal, 166; Lower Ninth Ward, 9, 131–32, 134, 135, 138, 139–44, 145–51, 153–57, 167, 176, 187; New Orleans East, 142, 143; Orleans Avenue Canal, 166; Pontchartrain Park, 142; Port of New Orleans, 52, 131, 140, 186, 187; Seventeenth Street Canal, 166; St. Bernard Parish, 139, 142, 186; St. Charles Avenue, 132; St. Claude Avenue, 144, 145; Uptown (neighborhood), 132. *See also* bayous

New Orleans Auxiliary Sanitary Association, 42, 43

New Orleans Board of Health, 42

New Orleans Board of Trade, 50, 51

New Orleans City Council, 146, 148

New Orleans Flood of 1849. *See* Mississippi River/New Orleans Flood of 1849

New Orleans Parish Medical Society. *See* Orleans Parish Medical Society

New York, 1, 22, 33, 62, 184, 192

New York Herald, 62

Nile River, 136

Nilson, Ellen Edwards, 65

Noah's flood, 64

non-native species, 2, 3, 8, 83, 85, 103–9, 111, 113, 116–22

North Carolina, 75, 90

nutria, 8, 111–14, 120

Nye, Bill, 3

Office of Recovery and Development Administration (New Orleans), 132, 149

Ohio River valley, 23

oil industry, 139, 143, 187

oil spill, 2, 3; BP oil spill, 3, 4, 10

Okeechobee Hurricane of 1928, 7, 8, 80–83, 85, 87, 90–94, 96–99, 137

Oliver-Smith, Anthony, 5, 11n3

Orlando, Fla., 96, 190

Orleans Levee Board, 169, 173

Orleans Levee District (OLD), 169, 171, 172

Orleans Parish Medical Society, 42, 52

Ousley, Clarence, 65, 68, 70

Pahokee, Fla., 89, 95

Palm Beach County, Fla., 86, 89, 94, 95, 97

Panama/Panama Canal, 37, 44, 47, 48, 49–51, 52, 53, 55, 89, 97, 190

Para, Brazil, 48

parasitology, 56

Pasteur, Louis, 40

Peard, Julyan G., 56

pelt farming, 112

Pennsylvania, 107

Percy, LeRoy, 114

Perrow, Charles, 161

pesticides, 113–14

Peters, Samuel, 19–21, 24, 25, 27, 30, 33

Philadelphia, Pa., 22

Philippines, 2, 47

Pickering, Andrew, 136, 141

pigs (feral), 8, 109–11, 114, 120

pirates, 80

plantations, 14, 17, 20–21, 29–30, 38, 42, 84, 85, 88–89, 90, 92–95, 99, 137, 138, 139

plasmodia parasites, 54

Platt Amendment of 1901, 97

political ecology, 133, 134, 135, 136–38, 140, 144, 148, 156

Porter, William T., 22

poverty. *See* social inequality

Pritchard, Sara B., 2

Progressive Era, 68, 72–73, 186, 189

public health, 37, 39–45, 47–56

Puerto Rico, 47, 48, 49, 51, 80, 88, 89, 91–93, 94, 97–98

Quarantelli, Enrico, 66, 77n21

quarantine, 6–7, 41, 42–44, 49, 50, 52

racism. *See* social inequality

railroads/railways, 18, 33, 38, 42, 67, 86, 88,

99, 115, 162; Florida East Coast Railway, 188

Ramos, André, 48

Reconstruction, 41

Red River, 18

Reed, Robert N., 117

Reed, Walter, 45, 46, 55. *See also* U.S. Yellow Fever Commission (Reed Board)

Reed Board. *See* U.S. Yellow Fever Commission (Reed Board)

resilience, 4, 6, 10, 64, 67, 121, 162, 179; meaning of, 12n15, 121

Ribas, Emilio, 48

Ricciardi, Anthony, 104

rice crops, 111

Rio de Janeiro, 48

rising sea levels. *See* sea-level rise

risk, 2, 3, 4, 6, 9, 10, 80, 81, 85, 87, 90, 96, 99, 119, 122, 132, 137, 138, 140, 141, 144, 161–65, 168, 175, 176–79, 187, 191, 192; risk society, 163, 180n7; social production of risk, 162, 163, 177

Roaring Twenties, 86–87

Robertson, Pat, 5

Rockefeller, John D., 188

Ross, Ronald, 54

Rousseau, Jean-Jacques, 67

rumor, 51, 63, 70, 71

Saigon (Ho Chi Minh City), 50

Sanarelli, Giuseppe, 45

San Ciriaco hurricane, 91

San Felipe hurricane. *See* Okeechobee Hurricane of 1928

Sauvé, Pierre, 14, 17, 20, 21, 29, 30, 31, 33, 34

Savannah, Ga., 42

Save Our Wetlands, 172

Schleifstein, Mark, 161

Schnakenbourg, Christian, 91

School of Hygiene and Tropical Medicine, 53

Schwartz, Stuart, 82

Scurry, Thomas, 72

seafood industry, 3, 118

sea-level rise, 1, 2, 5, 8, 165, 173, 174, 179, 184–85, 188, 192. *See also* climate change

seawall, 73, 86. *See also* barrier (flood/water); dike; floodgate; floodwall; levee

segregation, 142–43, 147, 150

seismic activity. *See* earthquakes/seismic activity

September 11 attacks, 184

Serviço de Prophylaxia da Febre Amarella (Yellow Fever Prophylaxis Service), 48

Shanghai, China, 50

Shrader-Frechette, Kristin, 108

Simberloff, Daniel, 108

Simpson, Charles Torrey, 189

Singapore, 50

slavery/slaves, 22, 68, 81, 87, 88, 139, 146

slow disaster, 3, 8, 120

snakes, 14, 15, 116, 117, 118, 153. *See also* Burmese pythons

Snow, Skip, 117

social inequality, 1, 2, 3, 5, 7, 8, 15, 25, 54, 63, 64, 67, 69, 71, 75, 80, 81, 88, 91–95, 105, 131–32, 137, 141–44, 145, 147, 148, 149, 154, 156–58

Solnit, Rebecca, 64, 66, 71

South Africa, 118

South America, 56, 82, 113, 114

Southeast Louisiana Flood Protection Authority–East, 178

soybean industry, 111

Spain, 44, 101

Spanish-American War, 7, 44, 50, 55, 67, 89, 91

spatial history, 135

Spirit of the Times (New York), 22, 33

St. Lucia (island), 53

Standard Project Hurricane (SPH), 166–68, 169, 173, 174, 181n23

Steinberg, Ted, 1, 5, 16, 85, 149, 157

Sternberg, George, 44, 45, 47, 50

storm surge, 9, 17, 65, 67, 86, 139, 140, 141, 165, 166–67, 169, 171, 173, 174, 177, 182n23, 186, 187

Suez Canal, 50

sugar/sugarcane, 8, 38, 81, 84–85, 88–92, 93–99, 111, 137, 154, 190–91

Sumpter, Amy R., 5

Swamp Land Act of 1850, 33

Swift, Jonathan, 23, 29

Tabasco Company, 112

Tennessee, 41, 42

tetanus, 92

Texas, 1, 3, 49, 51, 62, 64, 70, 71, 72, 74, 75, 106, 109, 110, 113, 115, 166; cattle industry, 113. *See also* Galveston, Tex.

Texas State Volunteer Guard, 72

Tidwell, Mike, 161

Tierney, Kathleen, 176, 177
Times-Picayune (New Orleans), 20, 21, 22, 23, 173
Tin Can Tourists. *See* tourism
tornado, 2, 15, 62, 90, 103, 163
tourism, 3, 86, 94, 96, 99, 110, 191
Towner, Horace, 92
trauma, 7, 64–65, 74–75, 119
Trinidad, 53
tropical medicine, 47–48, 53–56
Tschinkel, Walter R., 114
tsunami, 2
tuberculosis, 56
Turner, Bob, 178
Tuttle, Julia, 188
Tyndall, John, 40
typhoid fever, 45, 92
typhoon, 2

United Nations International Strategy for Disaster Reduction (UNISDR), 11
U.S. Army Corps of Engineers, 9, 34, 83–84, 96, 131, 140, 141, 164–69, 171–77, 181n15, 183n51, 186, 190–91
U.S. Army Medical Board, 44, 45, 55
U.S. Congress, 42, 111, 113, 166, 167, 168, 169, 171, 172, 173, 174, 175, 176, 178, 186, 190
U.S. Department of Agriculture, 46, 111, 113, 116
U.S. Department of Commerce, 95
U.S. Geological Survey, 190
U.S. Military Academy, 45
U.S. Public Health and Marine Hospital Service, 42, 45, 47, 49, 51–53, 55, 59n36
U.S. Public Health Service, 47
U.S. Sugar Corporation, 190
U.S. Water Resources Council, 167
U.S. Weather Bureau, 70, 166
U.S. Yellow Fever Commission (Reed Board), 45, 47–51, 53. *See also* Reed, Walter; Walter Reed Army Medical Center
University of Louisiana (now Tulane University), 42
University of New Orleans, 172
urbanization/urban planning, 6, 8, 38, 39, 87, 114, 133, 134, 145, 156, 164, 189, 192

vandalism, 62, 66, 71
Vaughan, Diane, 175

Vaz, Domingo Pereira, 48
vigilante, 62, 64, 69, 71, 75
Virgin Islands, 91, 92, 93, 97
volcano, 2
Volga River, 185
vómito negro. See yellow fever
vulnerability, 1–6, 8–10, 13n15, 23, 84, 93, 120, 132–37, 141–44, 155–56, 161, 162, 164, 175, 176, 179, 184, 186, 192

Wallace, Alfred Russell, 107
Walter Reed Army Medical Center, 55. *See also* Reed, Walter; U.S. Yellow Fever Commission (Reed Board)
Washington D.C., 46, 49, 92
Watson, Hewett Cottrell, 107
Webb, James L. A., 61n74
Weekly Delta (New Orleans), 20, 21, 27, 29
West Africa, 56
West Indian storm of 1928. *See* Okeechobee Hurricane of 1928
West Indies, 53, 80–81, 94
West Palm Beach, Fla. *See* Palm Beach County, Fla.
wetlands, 2, 8, 83–86, 89, 93, 94, 109, 111, 112, 119–20, 133, 138, 140, 143, 144, 150–54, 184, 187–88, 190, 191, 192, 193n9; preservation, 172; restoration, 151–54
whiskey, 27, 69
White, E. J., 25
whitetail deer, 111
Willson, J. D., 116, 117
Wilson, E. O., 104
Wilson, Woodrow, 73
Woodward, C. Vann, 114
World War I, 81, 83, 98, 186
Worster, Donald, 192

xenophobia, 105, 108. *See also* nativism
Xiamen, China, 54

yellow fever, 2, 6, 7, 37–56, 59n30, 61n73–74

Zanzibar, Tanzania, 50
Zika virus, 6